Simulation Modeling for
Watershed Management

Springer

New York
Berlin
Heidelberg
Barcelona
Hong Kong
London
Milan
Paris
Singapore
Tokyo

James Westervelt

Simulation Modeling for Watershed Management

 Springer

James Westervelt
Department of Agriculture and Consumer Economics
University of Illinois
326 Mumford Hall
MC-710/1301 West Gregory Drive
Urbana, IL 61801
USA
westerve@uiuc.edu

Library of Congress Cataloging-in-Publication Data
Westervelt, James D.
   Simulation modeling for watershed management /James Westervelt.
     p. cm.
   Includes bibliographical references and index (p.).
   ISBN 0-387-98893-9 (alk. paper)
    1. Watershed Management—Computer simulation.   2. Geographic information systems.
   I. Title.
   TC409 .W47 2000
   627′.042—dc21   00-059576

Printed on acid-free paper.

Production managed by A. Orrantia; manufacturing supervised by Jacqui Ashri.
PDF file prepared by Best-set Typesetters Ltd., Hong Kong.
Printed and bound by Edwards Brothers, Inc., Ann Arbor, MI.
Printed in the United States of America.

9 8 7 6 5 4 3 2 1

ISBN 0-387-98893-9   SPIN 10733338

Springer-Verlag New York Berlin Heidelberg
A member of BertelsmannSpringer Science+Business Media GmbH

Preface

This text introduces watershed managers and students of watershed management to the growing role of spatially explicit simulation modeling in the evaluation of management alternatives. Simulation modeling has a rich tradition in watershed-based scientific research. It has also been used to assist in evaluating watershed management decisions that are associated with very significant economic trade-offs. Education-based watershed simulation modeling is also available and being used in the classroom. Finally, geographic information systems have become commonplace tools in the management of watersheds and natural resources and are associated with inexpensive and growing libraries of digital map databases. Geographic information systems were developed to help us understand the consequences of our management actions. This technology is excellent for capturing the state of land and watershed systems, but that is only half of the picture. To reach their full potential in predicting and evaluating alternative watershed management decisions, geographic information systems must be associated formally with knowledge about the dynamics of our land and watershed system. This book introduces you to management-based watershed simulation modeling as a natural extension of geographic information systems. Whether you are a professional watershed manager, an interested citizen perhaps participating on a local watershed management committee, or a student interested in watershed planning, this book will take you on a tour that will better prepare you to access and use available and emerging watershed management technologies.

This book clearly represents a collaborative, community effort. The motivation for the book comes from years of hands-on simulation modeling work championed by Bill Goran of the U.S. Army Corps of Engineers Engineering Research and Development Center (ERDC). The funding that allowed the book to be completed came from the Illinois Council on Food and Agriculture Research through one of its leaders, Sarahelen Thompson of the Department of Agricultural and Consumer Economics at the University of Illinois, Urbana-Champaign. The source material was developed over many decades by many individuals working fervently on watershed and landscape simulation model development. The soul of the book evolved through many wonderful conversations with Bruce Hannon, Department of Geography, and Lewis Hopkins, Department of Urban and Regional Planning, University of Illinois, Urbana-Champaign. The readability of the text was greatly enhanced by Gloria Wienke, ERDC. The book was made more sensible by Manette Messenger, who lives to bring smiles to the world. Writing time was made possible through the loving support of my wife and best friend, Eileen. Very helpful editorials were provided by Matthias Ruth, Department of CAS Geography, Boston University and an anonymous reviewer. I am thoroughly blessed to be part of this community, thank you all. Finally, warmest thanks to the staff of Springer-Verlag, particularly Janet Slobodien and Tony Orrantia for shepherding the effort through the final stages.

JAMES WESTERVELT

Contents

List of Figures

Introduction

Management of natural resources has always been a contentious endeavor that forces us to recognize and understand that our communities are filled with individuals with dramatically different worldviews and life objectives. Individuals have dramatically different ideas about the consequences of alternative watershed management scenarios. Arguments arising from our differences of opinion are resolved through public hearings, legislative processes, and the court system. Science can play a critical role in the refinement of our collective ability to predict management consequences on nature and economics. These scientifically envisioned consequences must then be filtered through public opinion where notions of goodness allow us to select among a set of options.

Science is a process that formalizes our understanding and knowledge. Informal knowledge (our understandings) is recast as hypotheses, which are then subjected to measurements and scrutinized by a community of scientists. Knowledge that survives this intense scrutiny is formally retained as part of the body of science. Management is a process that must artfully combine formal and informal knowledge. Watershed management is necessarily an imprecise activity that has associated with it the philosophy of "muddling through." To avoid making poor decisions with long-term impacts, many managers adopt a strategy called adaptive management. They move toward what appears to be a good plan without fully embracing the entire plan. If this works, they move even closer. They make slight changes in direction as feedback associated with recent activities is collected. The steps that a managing body is willing to take are associated, in part, with their level of certainty in their ability to predict consequences of these steps.

A science-based formalization technique that is almost ubiquitously embraced is the geographic information system (GIS), originally developed in the mid-1970s. The acceptance of the GIS establishes it as a necessary tool in the intelligent management of natural resources. It provides two fundamental capabilities. First, it allows us to formalize our ideas about the state of our landscape and watershed systems. Second, it allows us to overlay, analyze, and probe the digital maps that store the formal information. It formalizes only half of the important information needed for us to predict the consequences of alternative watershed management decisions: system state information. To this system state knowledge we must add understandings of the system dynamics. Watershed management groups that regularly use GIS currently provide informal ideas about the system dynamics. Now adopted as an invaluable watershed management tool, GIS must be matched with watershed simulation models to complete a multi-decade transition that dramatically improves watershed management through a formal capturing of knowledge about watershed system state and dynamics.

This book offers five basic messages:

1. Simulation modeling is becoming more common.
2. Education-based simulation modeling is accessible to anyone.
3. Models developed for scientific investigations can be useful, but are by themselves inadequate.
4. For serious watershed simulation modeling, experts and their models need to be tapped.
5. Rapid advances are underway that will yield increasingly powerful management-oriented simulation modeling tools for watershed systems.

Simulation modeling is becoming increasingly useful in our society. Children are given simulation-based games. Equipment, including jet airliners, is simulated at the design phase. Scientists regularly use simulation modeling to understand and predict in the fields of hydrology, ecology, economics, traffic control, manufacturing, and business management. A number of software companies offer open-ended simulation modeling programs that are accessible to anyone with a basic background in algebra. The number of liberal arts students graduating with some experience in simulation modeling is increasing. There is a very significant difference between simulation modeling developed to support scientists and modeling to support natural resource managers. Powerful simulation models developed by hydrologists, ecologists, and other academic disciplines are inadequate and often inappropriate for assisting in the evaluation of alternative watershed management options. When these models are appropriate, it is typically wise to leave the operation of such models to the scientific community. Model input requirements can be daunting, the user interfaces are usually poor, and the outputs difficult to interpret. With patience and close collaboration with the scientists, the models can be useful in comparing alternatives. Finally, expect that in the coming decades, watershed-based decision support systems will increasingly use simulation modeling. Government agencies are working hard to provide watershed management decisionmakers with powerful multidisciplinary simulation modeling tools. Commercial GIS vendors will also be offering a growing variety of simulation modeling tools in their future releases.

This book is divided into three parts. Part I, History, Theory, and Challenges, provides background information on simulation modeling. The first chapter looks at the recent history that has led to the enabling of local watershed management. Chapter 2 reviews the challenges associated with simulation modeling, addressing various concerns regularly raised with respect to the reasonableness of doing simulation modeling. Chapters 3 and 4 look at simulation modeling as developed and implemented by the ecological and hydrologic modeling communities, respectively.

Readers interested in what is currently available to support management-based watershed simulation modeling can safely skip directly to Part II, Choosing Models and Modeling Environments. This section reflects on single-discipline models (Chapter 5), multidiscipline models (Chapter 6), and software that facilitates the development of altogether new models (Chapter 7). Simulation modeling to evaluate management alternatives is often multidisciplinary and requires that local situations be captured in a simulation model. Chapter 8 provides a recipe for coordinating the development of new multidisciplinary simulation models. Chapters 9 and 10 explore how consequences of management alternatives can be illuminated and explored with simulation models. Chapter 11 explores the challenge of model error and uncertainty analysis. Chapter 12 offers guidelines that can be used to evaluate available models for meeting the needs of a particular watershed management decision-making process.

Finally, Part III, An Integrated Watershed Modeling and Simulation Future, offers a glimpse into the near future when management-oriented simulation modeling will become as common and as important as GIS in the 1990s. Design philosophies (Chapter 13) are followed by views of an integrated watershed modeling and simulation system from the perspectives of the land manager (Chapter 14), the simulation model developer (Chapter 15), and the system developer (Chapter 16). The three parts together cover the past, present, and future of watershed-based simulation modeling.

Part I
History, Theory, and Challenges

1
Characterizing Watershed and Landscape Management

For several decades, watershed simulation modeling has been an important tool in the domain of scientists. Locally led watershed management approaches require that local citizens and watershed managers become familiar with these tools. How has locally led watershed management developed recently? This chapter takes a brief look at the emergence of locally led watershed management and then looks at current roles of watershed management committees and the watershed management plans they create.

1.1 Locally Led Watershed Management

Watershed-oriented management approaches are becoming increasingly popular. It is common to regularly hear about watershed analysis, watershed health, watershed coordination, watershed management, watershed committees, and so forth. Although management of our lands by watershed is now becoming popular, watershed management as a field of study and as an approach to land management has been promoted by scientists for decades. We have traditionally managed land with respect to historical political boundaries—often drawing straight lines across heterogeneous landscapes, watersheds, and ecosystems. Management differences on either side of these straight lines can sometimes be seen clearly in satellite imagery when the political boundaries affect the ecological processes. Figure 1.1 is a satellite image of a boundary between a military installation in Texas and private lands. The darker half of the image (southwest) is forested within the military installation and the lighter half (northeast) is private cattle-grazing land. Amusingly, political boundaries can also be seen on soil maps as a result of two survey crews completing work on different counties at different times. As a result, soil types do not appear to flow naturally across county boundaries. Areas are bound together naturally via different mechanisms. At a continental scale, climate patterns and topography give rise to ecoregions, which themselves are mosaics of land patterns at finer scales. Water is a critical resource for all life. At the continental scale, water helps

FIGURE 1.1. Different land covers on either side of a political boundary

define the ecoregions through annual rainfall cycles. At a more local scale, this water is delivered via a combination of processes that include rain, overland flow, groundwater flow, and mass movement via streams and rivers. Through these processes, the character of the water changes as it interacts with the biophysical world. The activities in the area upstream from any organism help define the quality and character of the water that contributes to the health of that organism. Herein lies the primary impetus for management of natural resources by watershed. Adopting watersheds as management units is an acceptance of the importance of natural geography.

The availability and quality of water at any point on a landscape is based on the movement of rainwater through (and under) the landscape upstream from that point. Habitat health is directly related to water availability and quality, and wildlife population levels illuminate that health. Except for human habitations in arid areas that import water from rivers flowing hundreds of miles away, drinking water is extracted from local streams, lakes, and aquifers. The quality of this water is a function of upstream natural processes and human activities. Landowners and tenants are linked to upstream neighbors by what is added to the water upstream that flows through their water, to us, and then to our downstream neighbors.

Why is watershed management becoming so popular today? Basically, the increasing population has resulted in an amplification of the impacts of local populations on downstream water quality and flow rates. Historically, "The solution to pollution was dilution." As point source and nonpoint source pollution increases, the ability to dilute diminishes until serious downstream health and safety problems appear. The seriousness of a number of high-visibility downstream impacts resulted in the passing of the 1972 Federal Water Pollution Control Act, commonly called the Clean Water Act (CWA) (Adler et al. 1993). Its primary objective was to dramatically improve the waters of the United States. Amendments to the CWA in 1977 and the 1987 Water Quality Act reaffirmed and strengthened the original act. In the original act, two national goals were established: stop all pollutant discharge into U.S. waters by 1985, and achieve water quality levels in these waters that would safely support swimming and fishing. A comprehensive framework was established to organize the development of standards, tools, and financial assistance to be used in response to water pollution and water quality issues. Approaches include limiting the discharge of wastes into waterways through a national system of discharge permits, development and protection of wetlands and aquatic habitats, grant and subsidy programs for the creation and upgrade of sewage treatment plants, and requirements for the control of chemical spills.

Agricultural lands and activities associated with normal farming, including fertilizing, planting, plowing, harvesting, and creating farm ponds, are not covered by the CWA. To encourage environmentally responsible farming, programs have been developed to financially compensate landowners for managing their lands in environmentally friendly ways. One example is the federal Conservation Reserve Program (CRP)[1] and the associated federal–state developed Conservation Reserve Enhancement Program (CREP). Landowners volunteer to take land out of production by planting trees, establishing wildlife habitats, and creating erosion buffer strips, and for this the federal and state governments financially compensate them. Through the CWA, the federal government established its interest in the quality of the nation's waters and, therefore, the quality of water flowing off private properties. Individual landowners struggle to retain control over their properties, including their rights to farm or build as each owner deems reasonable. States, counties, water control districts, cities, and other government entities all represent stakeholders in water quality.

The U.S. Environmental Protection Agency (EPA) is responsible for enforcement of the CWA. In recent years, the EPA has started to tap into local clean water interests to help communities meet the requirements and goals of the CWA (Ficks 1997). This has involved the organization of locally led watershed management groups, funding and support for those groups, and a new ethic that puts scientists "on tap, not on top." Instead of confrontations between "big government" and local landowners, enhanced

---

[1] CRP—http://www.fsa.usda.gov/dafp/cepd/crep/crephome.htm.

discourse is now occurring among local organizations, individuals, and landowners. Funding and support are available from various federal and state government agencies to create and apply locally developed watershed management plans. Through this support, ad hoc watershed management groups gain the necessary legitimacy required to make progress.

The EPA's programs to support locally led planning[2] include education, historical and current data on watersheds, guidelines, and financial and technical resources. Printed material is now augmented with powerful and extensive Internet resources. Education comes in the form of games for children and curricula for elementary and high school teachers. Many of the EPA's numerous databases, including toxic discharges to air and water, stream flow and quality information, and sources and quality of drinking water, can be searched by watershed and retrieved. Any interested citizen can readily access maps, charts, tables, text, and images associated with their local watershed. One of the key data repositories for stream and river flow and quality data is the Storage and Retrieval (STORET)[3] system database. This vast database is fully open to public access. It contains "raw biological, chemical, and physical data on surface and ground water collected by federal, state and local agencies, Indian Tribes, volunteer groups, academics, and others."

State and federal government agencies with water management responsibilities are relying heavily on local citizen-based watershed management groups to develop watershed management plans using academic and government agency expertise. To that end, advice, data, support, funding, and other resources are readily available to citizen groups through publications and electronic media, especially the Internet. Much of what is accomplished within many watersheds is due to the dedicated efforts of locally led watershed management groups, in partnership with their elected officials.

1.2 Management Committees and Plans

Watershed management committees are formed locally to create and implement watershed management plans. As discussed and described earlier, watershed management committees derive support from the availability of federal and state funding to create and implement watershed management plans. The EPA, the U.S. Army Corps of Engineers, and the U.S. Department of the Interior provide expertise and funding for the development and implementation of plans. The Natural Resources Conservation Service

---

[2] Surf Your Watershed—http://www.epa.gov/surf/.
[3] STORET—http://www.epa.gov/owowwtr1/STORET/.

(NRCS) promotes watershed management, with its regional offices often playing an important local role. Watershed management requires the voluntary cooperation and coordination of local interests in a watershed. Through consensus building, watershed management plans are developed to address issues of mutual concern to the various local, regional, state, and federal stakeholders. There has been a growing acceptance of watershed management groups; they are supported in recent legislation and in the speeches of politicians. The White House Council on Environmental Quality recently initiated the American Heritage Rivers program.[4] Locally led watershed management groups nominated more than a hundred rivers in the United States for heritage river status—far beyond the expectations of the White House. The CWA mentions watershed management groups and their role in meeting national water standards. Vice President Al Gore directed assorted federal administrators to develop a plan to realize the CWA mandate for "fishable and swimmable" waters. Under the leadership of the Secretary of the Department of Agriculture, a plan was prepared and then presented in the 1998 State of the Union Address as President Clinton's Clean Water Initiative.

Watershed management groups can significantly influence the management of local natural resources and can access federal and state funding and expertise to develop and implement watershed management plans. While the funding remains in the hands of the various government agencies, local citizens through watershed management groups increasingly influence the application of that funding. At a gross level, a two-step process is involved: first, create the management plan and, second, implement the plan. Funding to implement watershed management requires that a plan be completed. Plan development is not easy, but there are now many lessons learned (Ficks 1997) and a growing number of case studies and stories that will help new groups maneuver around known obstacles in the path to a plan (Heathcote 1998). Watershed management groups can control the application of watershed simulation modeling for the development, implementation, and monitoring of a plan.

During the past couple of decades, Americans have chosen more and more to organize themselves around clean water issues—especially local clean water issues. Government agencies have learned to work closely with local groups and their initiatives by offering expertise and funding. Progress can be very efficient when federal government agencies team up with local citizens groups. The federal EPA is leading the effort in making it easy to access expertise, funding, and vast databases.

Management of watersheds is firmly placed in the hands of locally led watershed management groups. Government and university scientists no

---

[4] American Heritage Rivers—http://www.epa.gov/rivers/.

longer have the opportunity to lead management efforts, but are instead called upon to participate on technical committees. Watershed managers and citizens participating on watershed management committees must become familiar with the scientific community's simulation models and participate in their application. The following chapters will help enable these managers and citizens to do so.

2
Challenges of Management and Modeling

To date, watershed management groups do not generally use simulation modeling. This chapter explores general ideas about modeling, how modeling is related to data collection, and under what conditions modeling can be cost effective. Justification for not using a model in a particular situation may be found here, but the chapter also encourages exploration of potentials of simulation modeling.

Consider that there are two fundamental questions that need to be addressed with respect to a management decision. First, what will be the consequences of that decision with respect to the human, economic, and natural systems? Second, how do we individually and as a community feel about those consequences? Simulation modeling can help us formalize scientific knowledge to address the first question. Through the democratic political processes, a diverse community addresses the second question via individual and group introspection. This book is concerned with the development and adoption of tools to help address only the first question. Accurate prediction of consequences allows for appropriate debates on alternatives.

Under what conditions and in what situations should simulation models be used in the actual management of natural resources? The challenges to simulation modeling are many and various: whether or not the design and development of models is appropriate; whether or not they are cost effective; even whether they are productive. You may have taken part in some of these debates and found yourself at times for or against models or modeling. The design and development of models can be extraordinarily expensive. However, making mistakes in land management can have disastrous and expensive consequences. In this chapter, different reasons and objectives for modeling are considered along with the pros and cons associated with landscape simulation modeling.

People often react with a significant amount of distrust to the idea of using simulation models for evaluating land management alternatives. Many claim that modeling has been tried and always fails. It is a waste of time, money, and human resources. And in reality, there have been a good

9

number of individual models that have failed to live up to full expectations. In fact, expectations can easily be too aggressively set. Modelers must avoid believing that their models are as "real" as the system being modeled.

Modeling is a basic human activity that we regularly use to make sense of the world around us. Each person formulates understandings, or models, of the people and things around us, and the interrelationships among those things. We cannot choose to not model. For example, when you make your way to work each day, you use internal conceptual models of your surroundings and the experience of getting to work. These models allow you to predict the intentions of the other travelers, the size and mass of their vehicles, the ability of the road or path to sustain your weight, and the behavior of the objects and other people sharing your space. If your model is accurate, you generally stop and go in a manner that gets you to work without incident. If your model is in error however, you can find yourself in serious trouble. Your model can be deadly wrong if the road beneath you suddenly turns unexpectedly icy, another traveler behaves in a reckless way, a violent storm suddenly appears, or another event disrupts your expectations. Based on models of the world, its state, and the rules by which it operates, each person "expects" events to unfold in a particular way. Young men are saddled with relatively high insurance rates. Their models might assume that their reaction times would keep them free from trouble. They might presume that all other drivers go slowly enough to be passed. Their models are not quite predictive enough to keep them out of trouble, and as a group, they have high insurance rates. With experience (that often involves accidents—some serious) the models are improved and the driving speeds attenuate to reasonable levels. But even the most experienced and practiced driver can be surprised with new and unexpected conditions.

For our models to be useful, it is important to ensure that the models of others are in agreement with our own. Culture, laws, and ethics provide a common framework for ensuring that this happens. Each academic discipline, for example, has its own developed culture and language that contribute to the sharing of the models. Each of us has models of standard individuals, situations, homes, cities, and ourselves. Our unconscious mind is continually seeking to match experience-based sensory input collected through our lives with these models. We make decisions based on our understandings (our models) of the world around us. Modeling is human, natural, and inescapable.

The role of modeling is changing in the workplace as more of our informal conceptual models are being recast as formalized models. The next three figures and associated discussions paint a picture of the past, present, and future roles of modeling in the workplace. Figure 2.1 depicts how a complex land management decision might be made today. Each person involved in the decision possesses one or more conceptual models of how the landscape works. Each rectangular box represents a single individual. Idea models are based on such things as formal academic training, experi-

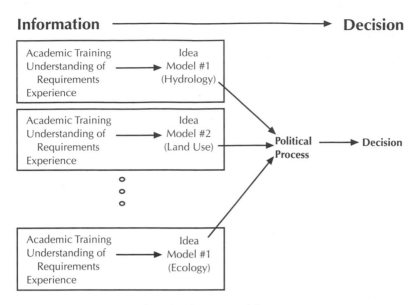

Figure 2.1. Current approach to landscape modeling

ence, and one's innate ability to work with abstract concepts and models. Typically, a number of individuals are involved. Each has a unique background associated with educational, professional, and personal experiences. These different backgrounds, combined with unique personalities, result in different understandings (models) of the watershed. Each person has different understandings of species requirements, habitat suitability, hydrology, biodiversity, genetics, chemical and noise impacts, and so forth. A potential management action posed to the team is actually posed to the team members' individual conceptual models. Based on the informal individual models, each team member can voice an opinion about the consequences of the management action. Typically, the opinions, based on different backgrounds, expertise, and personalities, are different. A political process is usually required to resolve the differences of opinion regarding how the human, economic, and natural systems will respond to proposed management strategies.

This approach to landscape modeling works well and has been the mainstay of complex multidisciplinary decision making in democratic societies. There are, however, a number of drawbacks to this approach. First, individual models are necessarily incomplete. Each individual is in possession of only a small part of all the information that is associated with the problem. Because no one person typically possesses sufficient knowledge about the problem to be able to provide an optimal answer, a team of individuals must cooperate, each representing a different academic background and field of experiences. Second, the models themselves cannot easily be evaluated because they are difficult to communicate. Typically, each individual is not fully aware of the complexity of the models they are using, if indeed they are even aware they are using a model. Idea models are not, in practice, clean logical inference models, but rather pattern-matching models that are associated with our ability to informally interpolate between patterns. Each of us looks at each new experience and tries to match it with similar experiences we've had in the past and generalize from old experiences to decide how we should behave in the new experience. Culturally we speak with logic, but in our minds we think with patterns. To communicate logic efficiently we must translate between the pattern-matching processes of thought and the formal logic of speech—a task that must be learned. Only after the models are communicated can they be easily evaluated. Third, since models are not easily communicated, combining different people's models is even more difficult and time consuming. Combining models is important when feedback loops among models are significant. For example, the hydrology affects the soil moisture, which affects the land-use patterns, which affects the hydrology. Multidisciplinary combinations of models can be potent—especially in complex feedback situations.

Today, the process outlined in Fig. 2.1 is beginning to change with the design and development of simulation models that represent one or more aspects of landscape dynamics. Figure 2.2 reflects a current trend of converting idea models into formal computer simulation software. In this figure, two of the idea models of Fig. 2.1 are shown captured formally in

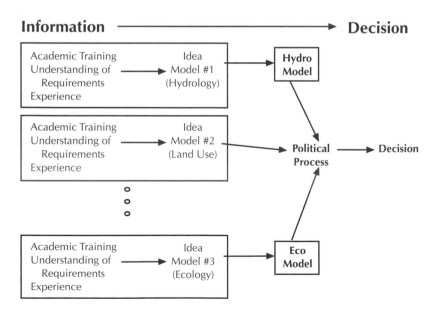

FIGURE 2.2. Emerging approach to landscape modeling

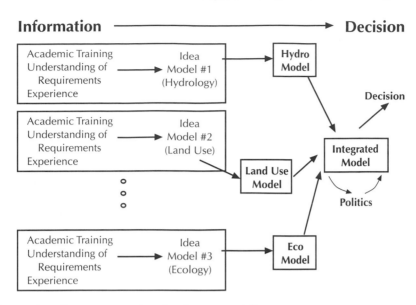

Information ──────────────────────────▶ **Decision**

FIGURE 2.3. Future approach to landscape modeling

software. The individual or group "owning" idea model #1 (hydrology) has captured their informal understandings of the hydrologic processes into a stand-alone hydrologic (hydro) model. So has the individual or group associated with the ecology ideas (model #3). Now, (1) more information can be associated with the model than a single person can capture, (2) the model can be evaluated and studied, and (3) it can be extended and attached to other models. This emerging approach to landscape management still requires a political process to integrate the output of the individual models with each other and with the remaining "output" from the conceptual models. Future management-oriented simulation modeling software will seek to transition from integrating human pattern-matching conceptual models toward using rigorous computer-based simulation models to support decision making.

Figure 2.3 suggests that the missing step required to complete the full transformation from conceptual modeling to computer-based simulation modeling is a collaborative modeling environment within which all professionals associated with landscape management can create integrated spatiotemporal ecological models. The separate hydrology, plant succession, land activities, habitat suitability, and other models can be designed and developed in a manner that allows each model to run landscape simulations in synchrony with the others.

Political processes continue to be important in the suggested series of management models described in Figs 2.1 to 2.3. In particular, formalized models cannot replace the values that individuals bring to decision-making processes. Formalized models only replace idea models for predicting the consequences of alternative management practices on our natural and economic systems. The value, or importance, of those consequences must still be debated among affected and concerned individuals.

Part III of this book develops a conceptual scheme describing the basic characteristics of this type of management modeling from the perspectives of the end user, the modeler, and the software developer.

2.1 Simulation Modeling Objectives

Clearly stated and understood objectives are essential ingredients of communication. Models developed for one set of objectives should not be used or evaluated with respect to a different set of objectives. Doing so results in the risk of misleading conclusions, poor management decisions, and poor reviews of the associated models. In this section, we look at different objectives for developing simulation models.

Most models can be grouped into two distinct categories: scientific understandings and management support. Most models encountered in the literature should be placed into the former category. This book, however, focuses on simulation modeling efforts developed in direct support of land management. We will first look at research-based simulation models.

Simulation Modeling for Scientific Understanding

Scientists seeking to understand nature are inclined to develop models of nature that embody their testable hypotheses and are then adjusted as the hypotheses are tested. The resulting models are typically difficult for others, especially individuals outside the discipline of the developer, to operate. The models are always discipline-centric. For example, hydrologists develop hydrology models associated, perhaps, with simple ecology to capture land-cover characteristics. Ecologists develop powerful ecological models that might be associated with very simple hydrologic models.

Natural systems operate with a large array of environmental variables in constant flux. To understand a component of the system it is necessary to focus on a small number of variables and evaluate how the system responds to changes in those variables. Scientific analysis, therefore, is done by artificially holding most natural variables constant. Consider a forest ecosystem. A small group of scientists might rapidly identify thousands of different measurements that could be taken from that system. A sample set of these is listed here:

- Chemical concentrations
  Within organisms
  In soils and litter
  In the atmosphere

- Species
- Population characteristics
  Age structure
  Counts
  Distribution
- Climate and weather characteristics
  Temperature
  Humidity
  Pressure
  Precipitation
  Lightning
- Geophysical characteristics
  Geology
  Slope, elevation, aspect
  Latitude
- Externalities
  Solar radiation
  Water flow into system
  Infestations
  Fire

Additionally, these characteristics can be measured at different scales in space and time and with different aggregations. They can be measured as averages over a particular combination of time and space or they can be measurements at points in time and space. The environment within which each organism or population interacts is different from the environment associated with any other organism or population. The complexity of a natural environment is significant. To understand the environment, scientists often begin by isolating a single variable, or a minimal set of variables. For example, a study might endeavor to understand the impact of nitrogen concentrations on vegetative productivity. To accomplish this, the study team might seek to hold the system steady by fixing the variables associated with the above list in a laboratory setting. The system variability must be held sufficiently constant to enable the study team to see correlation between the nitrogen and the productivity. This results in important information that can be published in the scientific record. Scientists attempt to draw cause–effect relationships from the data, thereby producing a model of what is happening in a small part of the system. This model might be a simple linear regression with associated statistics, or it may be developed as complex, time-variant, and nonlinear relationships. Ideally, the model can then be generalized to predict results not directly measured in the current experiments and in different environmental situations. New experiments are developed to test and validate the new model. Success is declared when a model is proven to have good predictive capabilities.

Simulation Modeling for Land Management

Models developed as a result of scientific studies are usually not immediately applicable to a land management decision or problem. There are two main reasons for this. First, the scientific studies strive to hold most of the environmental parameters constant while a small number of variables are carefully studied. Land managers cannot control a natural system as tightly as a scientific research team controls a relatively small experimental environment. Second, the temporal considerations in scientific studies, restricted by research budgets that span one to several years, are insufficient to support long-term predictions on complex landscapes defined by conditions that change slowly over time. Today, land managers are asked to make professional decisions based on a combination of training, experience, and analysis of published scientific results. The published studies might provide little more than anecdotal information to the land manager. The land manager knows that landscapes do vary in time and space and have different mixtures of plants and animals from those used in the scientific studies. Individual scientific studies, especially those undertaken in laboratory situations, must, unfortunately, be considered little more than suggestive.

Although results from a single scientific study may provide very limited information for making good management decisions, a rich historical scientific record reflecting changes to the watershed over years, decades, and longer, can provide significant information for making these decisions. Finding and then integrating the salient knowledge is time consuming and expensive. Today, information in the scientific record is combined through a mixture of land and watershed management experience, education, and discourse among interested and knowledgeable parties. Education consists of the formal training that professionals receive through school, reading of literature, and attending seminars, conferences, and workshops. On-the-job experience adds to the knowledge base of the professional. Watershed management is the art of combining all of this background to predict how alternative management scenarios are likely to affect the future status of the landscape, including its land patterns, water quality, natural habitats, and economic benefits. The manager's background and experiences are combined into conceptual models. These are informal in that they are ideas (potentially complex sets of ideas) captured in the minds of the land managers. Though informal, these conceptual models are very powerful, as they provide the substantive basis for decision making today.

Prediction can involve formal simulation modeling, which is the application of cause–effect relationships over time for the purpose of describing the future state of the system given a defined starting state and actions on and within that system over time. A formal simulation model is based on experience and facts and information formally published in books and articles. It is formalized by being written down as text or computer code that permits careful analysis and review by others. Unlike informal, personal conceptual models, formal models invite participation and comment; they become collaborative.

Each individual develops conceptual dynamic models of the world. The similarity of the models between any two individuals is based on the similarity of the experiences and education of those individuals. It is also based on the individuals' intellectual capacities, health, and underlying philosophies of life. A premise of this book is that no one individual has sufficient experience, background, and intelligence to make optimal landscape decisions. Multiple simulation models designed, developed, and used informally in the minds of professionals must, therefore, be communicated and combined. Traditionally, this process occurs through meetings, hearings, formal and informal discourse, and the political processes. Models (ideas) are presented, argued for and against, and contrasted and compared with other models.

Using the computer, the traditional combination of simulation models through discourse and politics is being augmented with the formal capture and integration of simulation models in software. Table 2-1 presents a simple seven-step conceptual flow of information that results in land man-

TABLE 2.1. Steps in making land management decisions.

Activity/Step	Traditional	Emerging	Future
1. Basic research	Captured in refereed journal articles	Captured in refereed journal articles	Captured in refereed journal articles
2. Training of professionals	Higher education	Higher education	Higher education
3. Model development	Conceptual models	Conceptual models	Conceptual models
4. Formalization of models	None	Discipline-specific computer models	Discipline-specific models
5. Formal model integration	None	None	Convert models to modules and integrate into single model
6. Impact prediction	Based on verbal debate of conceptual models	Based on verbal debate of discipline-specific conceptual and computer models	Based on running simulations on collaborative interdisciplinary models
7. Decision	Based on community values with respect to the predicted outcomes	Based on community values with respect to the predicted outcomes	Based on community values with respect to the predicted outcomes

agement decisions. The three columns represent this flow for "traditional," "emerging," and "future" approaches. The first three steps are identical in each approach. Scientific research results in the publication of new information that is formally presented through programs in higher education and in technical journals that are available to professionals. The knowledge gained by each individual results in an understanding (a conceptual model) of the landscape system (step 3). In the "traditional" approach, these understandings result in best professional judgments that can be presented by each individual involved in the management of the landscape. Differences of opinion are sorted out through verbal debate in formal and informal settings among the participants. Today, many of the conceptual models are being captured in discipline-specific computer simulation models. Formalization of models is an important step (step 4) that provides the advantage of formalizing the conceptual models in mathematical/logical models. When captured on a computer, formal models can easily be inspected, evaluated, and changed. The results from an assorted set of discipline-centered simulations (hydrology, ecology, biodiversity, economics, etc.) must then still be integrated through formal and informal communications among the participants in the landscape management decision. Predictions from a number of different academic discipline-centric models can easily be inconsistent.

The future approach adds step 5. Individual discipline-specific models are integrated into a single collaborative multidisciplinary model. Verbal debate focused on the interactions among the different models is now eliminated. Decisions regarding landscape management alternatives can be tested directly on the integrated model. The results of the tests can then be associated with the management objectives and trade-off values of the community. Communication associated with particular land management decisions can now ignore the meaning of inconsistent results from a number of different models and instead focus on the formulations of goals, objectives, and trade-off values.

2.2 Which Comes First, Data or Models?

How are data and models related? This section describes models and data as two inextricable parts of a whole. Neither is complete without the other. Some argue that models are useless without data and, therefore, should not be developed until there are sufficient data to support them. Others argue that data collection is efficient only when a developed model requires the data. The formalization of models is frequently associated with data requirements that often cannot be immediately fulfilled. But we must recognize that individuals are already using the associated conceptual models, without the supporting data. Formalization of the model reveals the importance of the missing information, which is a benefit of model formalization rather than a problem with formal modeling. A formalized model can be evaluated through sensitivity analysis of variables to identify the relative importance of missing data. Results can be used to cost-effectively select data to be collected.

Data collection schemes developed using models to guide and justify the collection may not be cost effective. The process of model sensitivity analysis helps to focus data collection on the most relevant information. This analysis determines how sensitive the model results are to slight modifications in model components such as equations and variables. Components with lower sensitivity can be more safely ignored in model creation than can other components. If modeling follows data collection, sensitivity analysis can reveal that while some collected data are important to model accuracy, other data are not, and still other required data were never collected. Models, including simulation models, are developed to address a particular question or set of questions. During the design, construction, and operation, sensitivity analysis on the importance of the variables should be conducted. During the early design of model development, many variables are considered for inclusion in the model. Some can then be disregarded if they are considered to be much less important than other variables. This is a conceptual sensitivity analysis. During model development, other variables are often removed as the team faces the constraints of time, expertise, and funding. Finally, after the model is constructed, a formal analysis of the importance of different variables can be conducted. It may be determined that certain variables are relatively insensitive and that a ballpark figure works as well as a more precise value. Other variables will be very sensitive. It will be cost effective to measure these variables through more precise field and laboratory research.

The collection of information about a watershed, the research and development of the watershed systems, and models of the landscape should be developed synchronously, evolving together. It is unreasonable to believe that a data collection effort can be designed in a manner that will ensure that the most important information is collected to support a future modeling effort. It is equally unreasonable to develop models without regard to data already collected or data collection projects already funded. Watersheds are extremely complicated and concepts, ideas, and theories developed to predict the future of the landscape abound. Each results in a different model that has different data collection needs. Watershed modeling and the collection of watershed state information must occur in concert.

2.3 Making Modeling Cost Effective

A number of different challenges are associated with the goal of making landscape simulation modeling cost effective. Challenges can be grouped into the following categories, discussed below:

- Human collaboration—how to accommodate multistakeholder participation?
- Software—what still needs to be developed?
- Defining watershed states—how to most effectively use geographic information systems (GISs)?
- Defining watershed processes—how to allow managers to define processes?
- Error propagation and sensitivity analyses—how to take modeling errors into account?

Human collaboration is perhaps the most interesting and surprising challenge in the above list. As discussed previously in section 2.1, most existing simulation models were developed for the purpose of understanding scientific principles operating within the landscape. However, these simulations and the knowledge and literature generated in association with the models have limited value to the land manager. Such information must be carefully combined with a large amount of other information developed through research in a large number of different areas, including biology, ecology, economics, agriculture, and others. Geography and regional planning are disciplines that lie at the interface or conjunction of these different fields. It is a very rare professional who can fully understand and then appropriately combine the scientific results published across a number of different academic fields. It becomes necessary to develop teams of individuals that represent the range of relevant disciplines. The development of future landscape simulation models will employ scientists, academicians, and practitioners from a number of different backgrounds. Procedures and processes that allow a disparate group of individuals to collaborate on the development of future models are being created and improved.

Management-oriented simulation models that allow watershed-oriented models developed in different disciplines to fully interact with each other are not yet ready for general use. Plenty of prototype environments do exist, but they have been developed for particular users, locations, and challenges (Trame et al. 1997, DeAngelis et al. 1998). Although there are good examples of two different simulation models being interfaced with one another, there are very few attempts to design software environments that allow for the integration of more than a handful of different models. One recent attempt is the Modular Modeling Language developed by the U.S. Geological Survey (USGS) (Leavesley et al. 1995).

Measurement of the state of landscapes and the components that make up landscapes is fundamentally important to the development of effective landscape simulation models. For most natural systems, good programs are in place to address this requirement for measurement. On-the-ground field efforts have always been important and are now being augmented with remote sensing techniques. Remote sensing provides a very small amount of information about any point on the landscape, but provides full coverage. The combination of field studies that collect very rich data sets for a few sites, with remotely sensed images that collect a little information for every 10- to 50-m^2 patch on the landscape, can be very powerful. Since the early 1980s, digital maps have been developed at a variety of scales across the world. Many of these have been based on the processing of remotely sensed imagery supported by field research. We have fairly well-developed programs that have given us a picture of the state of our landscapes. The programs include soil surveys conducted by the Natural Resources Conservation Service (NRCS), digital elevation maps developed by the USGS, aerial photographs coordinated by USGS, satellite imagery streaming from a number of platforms, and various state and local programs.

The natural processes that drive our landscapes and their associated ecosystems are relatively poorly understood. While measurements of the state of a given system are relatively inexpensive, can be accomplished in a short amount of time, and can be expressed statistically, understanding the processes that drive natural systems needs further investigation. It is these processes that we want to capture in our models. Chapter 3 explores current ecological theories and understanding of how natural systems are organized and how they work. General theories have been established and can be useful for predicting landscape and watershed responses at gross scales.

The final challenge facing developers of landscape simulation models designed to support land managers is the question of error propagation and sensitivity analysis. Virtually every aspect of a landscape simulation model is associated with error. Errors are involved with the conceptual design, measurements of the landscape used to initialize the model, estimation of various parameters, algorithms used to define the model, and even computations executed at the hardware level of the computer. The importance of certain errors is established through a procedure called sensitivity analysis. This procedure also provides a land manager with an understanding of how important it is to control various aspects of the landscape. Chapter 10 reviews errors, their sources, and how they might be tracked through a simulation.

2.4 How Formal Should a Model Be?

If modeling is inescapable, how can we choose the best model for a given situation? What criteria should be on the checklist used to evaluate the utility of a model? Two fundamental criteria should be applied: (1) what is the cost of developing and maintaining the model with respect to the benefit, and (2) under what conditions is the model appropriately accurate? To help with a cost–benefit analysis, let's look at a model taxonomy organized by cost. The hierarchy we will use is this:

- Commonsense models
- "Rule of thumb" models

- Expert models
- Scientific models
- Multidisciplinary management models

Generally, as we proceed down the list, cost can be expected to increase. Hopefully, accuracy increases also. A short exploration of each approach will help illuminate the pros and cons of each.

Commonsense Models

For this discussion, commonsense models are the conceptual models that we all carry with us. They are our informal understanding of how the world works and they allow us to choose courses of action. As we get older, our models become more finely tuned through the trial and error of experience. We do not automatically develop identical commonsense models and we all find ourselves musing how certain others have managed to survive so long with their "crazy" ideas. Language allows us to communicate, with difficulty, our different models to one another—sometimes resulting in surprise and shock. Common sense, we often find, is not commonly shared. Each of us develops a slightly different model of our shared space. If these models get us across the street, allow us to build careers, help us to communicate with others, and are sufficiently accurate predictors of future consequences of our actions, then we get along quite well with the world.

Our "commonsense" models of the world are extremely multidisciplinary. Crossing the street involves complex considerations about the physics of moving vehicles, psychology of the local drivers, and our own physical capabilities. Beneath this level is the complex coordination of our muscles and the associated neural processes. Imagine programming a computer to cross roads and it is easy to appreciate the power and complexity of our standard get-through-life models. In a watershed setting, our models associate rain with river levels, seasons with vegetation, clouds with expected weather, bare ground with silt loads, and location of construction with flooding risk. Our models are very inexpensive from the standpoint of marginal cost—because they are ready to use at any time. Education (formal and informal) is the process that develops, tests, and improves models, and provides our commonsense models with data. Although extremely useful and immeasurably valuable, commonsense models lack a certain formality required for unambiguous communication among individuals. Sometimes our personal commonsense models can be reduced to a "rule of thumb."

"Rule-of-Thumb" Models

Commonsense models, which are easy to state and easy to accept, can become part of our common culture. We pass these among ourselves as a "rule-of-thumb" model. An engineer might design a bridge to accept anticipated loads and then triple the strength of the beams. A gardener might plant bulbs at a depth three times the bulb diameter in the south—four times in the north. A project bidder might estimate the expected time the project should take and then bid double that estimate. Consider crossing the street: if the light is red, don't walk. Often, complex scientific research and related models will yield a new rule. Many scientific studies and complex chemical pathway models back up the rule that moderate exercise decreases many health risks. "Rules of thumb" are very simple statements that make decision making easy. They do not provide reasons for the decisions, but they can be accurate and useful.

Expert Models

Expert models are similar to commonsense models in that they are the conceptual models that have developed through years of training, study, and practice in a particular discipline. A visit to a doctor, lawyer, teacher, or scientist puts us in touch with much more sophisticated and elaborate models of the world with respect to the discipline that individual represents. A degree or certification on the wall of the professional is recognition that the individual shares at least part of their model or view of the world with a governing board in that profession. A visit by a plumber, electrician, healthcare worker, or builder similarly puts us in touch with worldviews accepted and shared by their respective disciplines. These professionals all have highly developed expertise-based conceptual models of certain aspects of the world. We can tap into that expertise, for a price, at any time.

Scientists, engineers, and their expertise traditionally have been powerful forces in the management of watersheds, lakes, rivers, and streams. Governing bodies controlling vast stretches of land could tap scientists and engineers for the development of a management plan. Dams could be constructed, lakes created, swamps drained, and rivers rerouted—all based on the expertise embodied in formal professions. Professional expertise is always in flux because of new knowledge emerging from basic and applied research efforts. And it is possible to have competing and disagreeing expert models within any given profession.

Scientific Models

A scientific model is a formalization of an expert's conceptual model into a form that can be communicated directly to other experts. Such models can be expressed in text through refereed journal articles. They may also be expressed as formalized algorithms captured in computer instructions. In journal articles, other scientists can study, replicate, and thoroughly analyze the accuracy of the described understanding or model. In computer software, a computer can reflect back to the scientist the implications and consequences of the model under different constraints. Models that hold up to the scrutiny of other scientists become valuable currency from which new technical innovations are developed and change our world.

Formalized scientific models are less common than expert opinion because few opinions can hold up to scrutiny from a body of individuals. A formal model that has held up well to scrutiny can be useful in building consensus and informing the judgment of expert opinion.

Scientific models can be very useful to watershed management. Chapter 5 reviews a good number of single-discipline science-based models that have been developed to support watershed management. One serious challenge to such models is that each typically models some aspect of the watershed processes in some significant detail, but models other aspects in a very simple manner, and sometimes not at all. A hydrologic simulation model might hold the landscape characteristics, such as vegetation cover, constant while modeling the flow of water in great detail. A vegetation succession model, on the other hand, may change plant species densities over time based on detailed plant interactions with other plants, soil characteristics, and nutrients. However, it uses annual average soil-moisture indices and completely neglects storm-water runoff. Consequently, the hydrologic and vegetation succession models might lead to very different management approaches.

Multidisciplinary Management Models

In response to the shortcomings of single-discipline scientific models, many scientists are collaborating to build integrated multidisciplinary models in support of watershed management. Serious challenges to these efforts include the development of difficult standards for software interaction, the cost of integration, and the acceptance by watershed management groups of the veracity of the integrated models. Much of this book is dedicated to the development of an appreciation of the costs and benefits associated with multidisciplinary management models.

Which Modeling Approach is Best?

On a day-to-day basis, managers rely on the informal commonsense models, rule-of-thumb models, and expert models. The formal scientific and multidisciplinary models are used by scientists and can affect management decisions through refinement of a manager's informal modeling approaches. Determination of the appropriate modeling approach is based on a number of complex and interrelated issues. Funding, time, accuracy requirements, legal precedents, local expertise, stakeholder interest, scientific knowledge, and success of others all contribute to the decision. Funtowicz and Ravetz (1991) argue that two of these factors can be used to identify the appropriate decision-making level: stakeholder interest and scientific knowledge. If the stakes are low and the scientific uncertainty is low (i.e., the science is

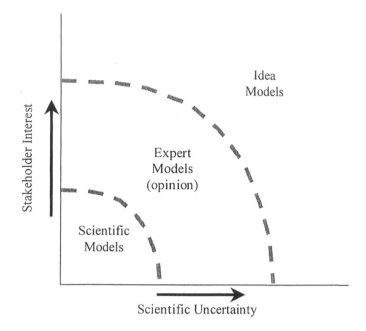

FIGURE 2.4. Role of modeling with respect to levels of stakeholder interest and scientific uncertainty

well understood), applied science is sufficient. If the stakes and/or the uncertainty are higher, then professional consultancy must replace the applied science. If either is higher still, the decision must be established through "postnormal science." "Postnormal" marks the passing of the time when focused science was believed philosophically, if not always practically, sufficient to address societal challenges. Funtowicz and Ravetz (1991) reflect on a new age in which science provides partial insights that must be enriched with local experience, extended peer-group insights, and appreciation for the beliefs and feelings of stakeholders. They define the following three continuous approaches to decision making: applied science, professional consultancy, and postnormal science. Decisions made at the wrong level may not be implemented, may be overly costly, and may need to be rethought. Parallels can be drawn between these approaches and associated modeling techniques. Figure 2.4 suggests that scientific models are adequate for the application of applied science in the face of low stakeholder interest and low uncertainty. Expert models support professional consultancy, and postnormal science is supported by idea models.

Each of these modeling paradigms or approaches can be valuable in land management. Scientific models can be sufficient if scientific certainty is high and if stakeholder interests are low. Stakeholders are not going to take the time to involve themselves due to lack of interest. If the uncertainty and/or the stakeholder interests are higher, science-based models are inadequate, though potentially still valuable, and expert opinions can be sufficient. Higher interest and/or higher uncertainty requires that the final decisions rest with the political processes. Here, science-based models can be used to support expert opinions that in turn are entered as testimony to the political process. This book argues that the role of scientific models can be significantly expanded through their integration into multidisciplinary management models.

3
Perspectives in Ecological Modeling and Simulation

Dynamic watershed simulation modeling must rest firmly on modern ecological theories. It is important that watershed modeling consider a broad array of current theories in ecology. A model can be inadequate if it embraces (and therefore facilitates) a single ecological theory. Wu and Loucks (1991) state: "A hierarchical perspective is appropriate and necessary to unify ecological concepts and theories. The unification can be accomplished only by focusing on the multiplicity of scales of ecological phenomena. Such a unifying perspective does not preclude, but builds upon, pluralistic studies at different ecological scales."

A literature review covering theories of ecology and past scientific accomplishments is presented below in section 3.1. The review includes equilibrium theory, nonequilibrium theory, hierarchy theory, metapopulation and patch theory, and landscape ecology. These theories and accomplishments provide the historical context within which ecological modeling and simulation should be accomplished. Then, section 3.2 identifies and briefly reviews some ecological modeling systems

3.1 Underlying Theories of Ecology

This section provides a brief literature review of current theories of ecology. These theories must be recognized when developing new ecologically based simulation models to support making management decisions. Each discussion briefly presents a current theory in ecology and then draws guidelines from this theory for the design and development of ecological modeling and simulation software. Conclusions from these theories are captured in Chapter 14 to help define a next-generation modeling environment.

Equilibrium Theory

Equilibrium theory assumes that a system, when perturbed, seeks to return to the equilibrium state. Wu (1994) reports that Plato and Aristotle pro-

vided the first supraorganismic balance-of-nature concept and Carolus Lin-
naeus (1707–1778) called the balance Oeconomia Naturae. Clements (1936)
put forth the organismic viewpoint of community ecology and advocated a
succession process that leads to a climax state—a state of natural equilibrium. Analytic approaches to modeling nature grew from this philosophy.
One of the first was Pierre-Francois Verhulst's logistic equation reviewed
in Kingsland (1985). This was joined by equations such as the Lotka–
Volterra, Rozenzweig–MacArthur, Leslie's predator–prey, the Nicholson
–Bailey model, and modern derivatives reviewed in DeAngelis and
Waterhouse (1987). Mostly analytical, these equations and models all have
well-defined equilibrium points. Even those that settle into a dynamic equi-
librium define a final steady state condition.

The basic logistic equation in difference equation form is

$$P_{T+1} = P_t + R*(C - P_t)/C \tag{3.1}$$

where P = population
R = maximum reproductive rate
C = carrying capacity

The population at each time step is the population of the previous time step
plus births. Births decrease as the population reaches the carrying capacity.
An example simulation output using the logistic equation is shown in
Fig. 3.1.

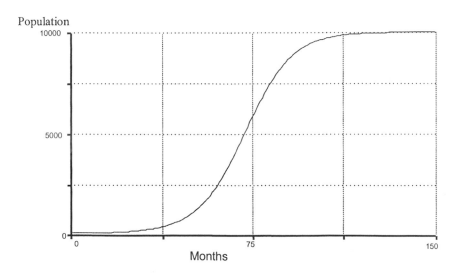

FIGURE 3.1. A logistic growth curve

The Lotka–Volterra model is a two-species predator–prey model expressed as differential equations. Let P represent the population of the prey and H the population of the hunters (predators). The basic equations are

$$dP/dt = a * P - b * P * H \tag{3.2}$$

$$dH/dt = e * b * P * H - c * H \tag{3.3}$$

where a = natural growth rate of prey (P) in absence of predation
      b = death rate of prey (P) when they encounter hunters (H)
      c = natural death rate of hunters (H) when prey (P) are unavailable
      e = efficiency of prey (P) into hunters (H)

dP/dt is read as the change in prey with respect to time. The "d" is short for "delta," traditionally used in mathematics to denote change. Equation 3.2 states that the change in prey (think prey mass) at any time is a natural constant growth rate of the prey (a) times the current prey population minus the natural death rate of prey when they encounter hunters. Note that increasing prey or hunters results in a greater death rate. The change in hunters (e.g. 3.3) is an efficiency of turning prey (mass) into hunter (mass) times the loss of prey due to interactions (in eq. 3.2) minus the natural death rate of the hunters. These equations form the foundation for specifying multiple predator–prey systems in a natural setting.

The basic Nicholson–Bailey host–parasite model, expressed as difference equations is

$$H_{t+1} = RH_t \exp(-aP_t) \tag{3.4}$$

$$P_{t+1} = cH_t[1 - \exp(-ap_t)] \tag{3.5}$$

where H = host population
      P = parasite population
      a − "effective area of search of the parasite"
      c = average number of parasites that emerge from an infected host
      R = intrinsic reproductive rate of the host (H)

$\exp(-a\,P_t)$ represents the proportion of hosts that survive parasitism based on the parasite search efficiency and density.

The Leslie models divide populations into age cohorts and specify survival rates for each age cohort to predict ongoing populations. Reproductive rates associated with each age cohort result in changing population birth rates over time. These three sample models (Nicholson–Bailey, Lotka–Voltera, and Leslie) are examples of many population and population interaction equations that have been used to describe and predict population changes over time. They represent a set of powerful yet simple models that are still used today to teach and understand population biology. For the most part, such equations define population trajectories that reach a steady state equilibrium.

If equilibrium states can exist in nature, it can be argued that these states are rarely, if ever, seen in practice because disturbances are continually pulling the system away from equilibrium. Following the disturbance, the system then continues to seek the equilibrium or climax state. Disturbance theory argues that ecological systems are continually in a state of flux. Analytic systems of equations are only useful for modeling systems that are at or near some equilibrium point (Reice 1994). Disturbance is also viewed as a major contributor to diversity, for through disturbance, additional ecological niches are opened for exploitation. Disturbance also occurs at various ecological scales (Pickett et al. 1989). Ecological systems from individuals to ecosystems can and are disturbed. Disturbance can be described and characterized by spatial distribution, frequency, return interval, rotation period, area, intensity, severity, and synergism.

Nonequilibrium Theory

Wiens et al. (1986) note that studying long-term natural processes is similar to creating an entire movie from a few frames. At human scales, it is easy to view nature at larger scales as systems in equilibrium, or at least systems that move toward equilibrium. However, it is not correct to assume that systems that appear to be stable at human scales are actually at equilibrium. Perhaps natural systems are actually in constant flux and equilibrium is never really attained. Wu and Loucks (1991) and many others recognized that equilibrium states require density-dependent population regulation and claim that there is little direct evidence for this. Caswell (1978) states, "Equilibrium theories are restricted to behavior at or near an equilibrium point, while non-equilibrium theories explicitly consider the transient behavior of the system." DeAngelis and Waterhouse (1987) write that the "dynamics of ecological systems at small spatial scales is usually an ephemeral phenomenon with no equilibrium properties." It is further argued that ecological systems at some scale demonstrate nonequilibrium processes that appear in equilibrium at larger scales (O'Neill et al. 1986, Urban et al. 1987). For example, fire is destabilizing at short time intervals, but stabilizing at long time intervals (Loucks 1970).

Nonequilibrium theories are grouped by Chesson and Case (1986) into four distinct types:

1. Those with an "absence of point equilibria." Such systems continually fluctuate in mathematically random and chaotic fashions. These systems can be argued to be in equilibrium in the sense that the possible states of the system are grouped around chaotic attractors; that is, the measured states of the system fluctuate (and sometimes oscillate) around one or more particular states, but never settle into equilibrium.
2. Those that emphasize "fluctuations in density or environmental variables as dominant processes."

3. Those in which the mean of climate fluctuations varies over time so that historical factors are important. Here, ecological systems adapt over centuries.
4. Those with competitive displacement. "Chance and history may be major factors shaping community structure." This is the expect-the-unexpected viewpoint. Events that appear to be random at a given scale may be a predictable part of a much larger time scale.

Essentially, nonequilibrium theories all agree that natural systems are in a continual state of flux that is occurring simultaneously at multiple scales.

Hierarchy Theory

The predictability of ecological systems is inherently limited and dependent on scales (Loucks et al. 1985, May 1986, Levin 1989). Any natural system can be studied at a variety of scales: from the subatomic to the ecosystem, and beyond. The degree to which any given ecological study identifies the existence or nonexistence of processes that allow that system, when perturbed, to return to some equilibrium state is dependent on the chosen scale. Different processes are seen and studied at different scales. "Therefore, there is no single correct scale of investigation and thus no universal law in ecology" (Wu and Loucks 1991).

Wiens et al. (1986) write, "Some of the most vociferous disagreements among ecologists arise from differences in their choice of scale." To illustrate the point, they suggest how differently ecologists studying the relationships between jackrabbits and coyotes at five different scales might view their interactions. These scales were defined as

1. The specific location where the animal lives; its home territory.
2. A local patch occupied by many individuals, the full spatial extent occupied by a population.
3. Many local populations that interrelate through dispersal; a metapopulation.
4. A closed system (or approximation thereof); an "island" of continuous physical and ecological characteristics separated from other similar "islands."
5. A biogeographical scale where different climates and different sets of species exist.

Depending on the spatiotemporal scale chosen, the two species can appear to be highly interrelated or completely independent. A short-term behavioral study at a single site will discern the feeding habits of the jackrabbits and coyotes. Interactions between the two species may or may not occur during the study, perhaps depending on the season. Conclusions about the interspecies interactions would be difficult to ascertain. A short-term field experiment that tracks population changes under controlled situations might conclude that the predator–prey interactions are crucial to both species. A medium-term field experiment might identify that although the two species do interact in a predator–prey relationship, their populations are not dependent on the existence of each other. A medium-term census over 10 years might not identify any particular relationship between the species and may determine that their populations are mostly affected by weather conditions. A short-term regional census may find negative or positive correlations between coyotes and jackrabbits, or may find no correlation, depending on the selection of sampling sites. Each independent view of the system has strengths and weaknesses, and each adds to a complete understanding. Land managers, modelers, and ecologists must always be willing to back away from their particular models and approaches and view the system from perspectives arising from different scales in time and space. This will ensure that the appropriate scales are chosen with respect to the particular question or set of questions being asked.

Hierarchy theory offers a framework within which to view and integrate different scales. The theory has matured sufficiently to be documented in several books (Allen and Starr 1982, O'Neill et al. 1989). There are three dimensions: time, space, and organization. Organization refers to organizational levels of life, which are often viewed as nested. Atoms are organized into molecules, molecules into cells, cells into organs, organs into individuals, individuals into populations, populations into communities, and communities into ecosystems. Natural phenomena that are represented by a large number of samples at the scale of study (e.g., atoms of an element in a sample or mice in a county) can be handled very well through statistical approaches and are called large-number systems. Phenomena that are represented by very few samples can be handled by careful and thorough study of each sample (low-number systems). Landscapes, when studied at the human scale, have too few components to treat statistically and too many components to study each thoroughly. Such "middle-number" systems are the focus of hierarchy theory (Allen and Starr 1982).

Hierarchy theory links these organizational levels. Lower levels operate in smaller partitions of space and shorter periods of time. Individuals operate on small scales in time and space, while ecosystems operate on relatively much larger scales in time and space. The apparently neat relationship among these three scales has been discussed and graphically depicted in time–space diagrams. Ocean hydrodynamics (Stommel 1963) and processes in landscape ecology (Urban et al. 1987) have been presented in such diagrams showing a clear and simple relationship (Johnson 1993). Delcourt and Delcourt (1991) partition time and space into four domains (overall time and space of interest):

- Microscale (1 to 500 yr, 1 to 10^6 m^2). This domain is the most familiar to ecologists. Within it exist population dynamics, productivity, competition, and response to disturbance events.

- Mesoscale (10^4 yr, 10^{10} m^2). Here landscape mosaics and watersheds dominate. Animals and plants develop adaptation to disturbance regimes.
- Macroscale (10^6 yr, 10^{12} m^2). This scale involves quaternary studies. Species displacements occur on a subcontinental scale, and rates of spread of species and genetics as well as extinctions define this scale.
- Megascale (>10^6 yr, >10^{12} m^2). At this scale, planetary phenomena, such as development of biosphere, lithosphere, hydrosphere, and atmosphere, and macroevolutionary history of life on earth dominate.

Systems, according to hierarchy theory, result from evolutionary processes that favor a nested, hierarchical organization. Each level is constructed from identifiable subsystems (Johnson 1993). For ecological modeling, the modeler need normally consider only three levels: the level dealing with the question being asked of the system, the next-higher level to provide context (constraints), and the next-lower level, which contains the dynamics and structure to be modeled (Johnson 1993). Dynamics of even-lower-level structures in the hierarchy are, for the most part, sufficiently attenuated to be replaced by average behaviors or even ignored because they are captured in an attenuated and aggregated fashion through dynamics occurring in intermediate levels (O'Neill et al. 1986). Landscape ecologists, for example, attempt to capture the complexities at smaller-than-landscape scales into single numbers and indices (Turner et al. 1989).

There is not always a strict hierarchy that proceeds nicely through atoms, molecules, organs, individuals, and so forth. Examples are ecosystems that exist within individual organisms (intestinal tracts), populations that can exist within individuals or other populations (diseases), or populations that span a number of communities (birds).

Metapopulation and Patch Theory

There is a notion of a spatial distribution of resources and processes. The theoretical underpinnings of this approach are found in the theories of island biogeography (MacArthur and Wilson 1967) and early arguments supporting notions of metapopulation theory (Andrewartha and Birch 1954, Hanski and Gilpin 1991). Island biogeography provided a theoretical and mathematical framework for describing the relationships between stable populations and nearby, but disconnected, areas of unstable populations (islands). Metapopulation theory extended the concepts of island biogeography by describing interactions among numerous areas (islands) containing unstable populations. It provided a theoretical foundation describing processes by which similar competitors can coexist in a patchy environment (Levins and Culver 1971, Horn and MacArthur 1972, Slatkin 1974).

Simple, but powerful, spatially explicit numerical models developed by Levins (Levins and Culver 1971) formed the foundation for a body of literature exploring metapopulation theory and its relationship to metacommunities, landscape ecology, island biogeography, patchy environments, and conservation biology. For a review, see Hanski and Gilpin (1991). Metapopulation theory provides a simple mechanism that explains how it is possible for a landscape to contain a number of direct competitors. In a completely homogeneous environment, the most successful competitors crowd out their inferior competition. Real systems are patchy at all levels of hierarchical organization because of perturbations and disturbances. In such dynamically heterogeneous environments, metapopulation theory predicts the existence of a potentially unlimited number of close competitors. Levins's basic equations have been extended in various different ways. Hanski (1985) added migration to Levins's model (to create a three-state model). Dynamic complications, caused by immigration, were demonstrated to result in alternative stable equilibria. Gilpin (1990) demonstrated numerical computer models for making predictions of the dynamics of real systems using metapopulation theory. Finally, Gardner et al. (1993) conducted theoretical simulations of competing species with varying perturbance regimes and harvest schemes.

Dispersal is the key activity driving metapopulation dynamics. Hansson (1991) reviews the characteristics of dispersal that influence metapopulation functioning and identifies these three categories of factors influencing dispersion: economic thresholds, resource conflicts, and inbreeding avoidance. Economic thresholds for the population involve the relative availability of essential environmental resources, such as food, shelter, and water. Resource conflicts often involve these resources as well as the conflicts over potential mates. Even when resources may be plentiful, territoriality behaviors may encourage migration. And some migratory behaviors seemed to be associated solely with drives aimed at avoiding inbreeding. Population size correlates with dispersion rates of different creatures.

With respect to hierarchies of system organization, Hanski and Gilpin (1991), in a historical account of metapopulation theory, define metapopulation scale as part of the following continuum:

- Local scale—the scale at which individuals move and interact with each other in the course of their routine feeding and breeding activities. This is the scale immediately beneath the metapopulation scale that provides the dynamics behind metapopulations.
- Metapopulation scale—the scale at which individuals infrequently move from one place (population) to another; typically across habitat types that are not suitable for their feeding and breeding activities, and often with substantial risk of failing to locate another suitable patch in which to settle.
- Geographic scale—the scale of species' entire geographical range; individuals have typically no possibility of moving to most parts of the range. This scale provides the context within which metapopulation dynamics take place.

While metapopulation theory provides the fundamental basis for accepting that multiple species competing for the same resources can coexist on a landscape, patch theory provides a framework for capturing the dynamics in spatially explicit models. The patchwork of landscapes and habitats has been shown to be very important for allowing local extinctions to occur, while maintaining populations at a broader scale (DeAngelis and Waterhouse 1987).

Tilman and Downing (1994) argue persuasively for the recognition of spatial arrangements as important variables in ecological models. They argue, from the literature, that colonization limitation is important in succession dynamics. Succession theory defines, for any given climate, a succession of vegetative communities. The fundamental basis of succession theory was initially laid out by Clements (1936). The actual trajectory of the succession of plants is not fixed, but is a function of the available seed sources; the local disturbance regimes as defined by fire, disease, flood, and other major perturbations; and the physiological capacity of the plants to respond to the perturbations.

There are numerous theoretical demonstrations that habitat subdivision allows two species to coexist as metapopulations, stabilizes host–parasite and predator–prey interactions, and influences the evolution of cooperative behavior (Tilman and Downing 1994). They present a spatially explicit model based on the work of Levins (1969) and the extensions by Hastings (1980) and Nee and May (1992). This model demonstrates that (1) a set of species cannot occupy all space in a simulated raster environment, (2) there can exist only a single individual representing a single species at any cell, (3) there is no spatial advantage given to any species for populating an open cell (species in adjacent cells have no advantage over other species), and (4) because competitively superior species are poorer distributors, any number of species can cohabitate. Using this model, Tilman and Downing (1994) also demonstrated that habitat destruction has the greatest effect on the best competitor.

Spatially explicit models using fixed grids of patches have been used to explain observations from nature. Reice (1994) demonstrated that when openings in a habitat are created, the proportions of the species that recolonize are unpredictable. Increased levels of disturbance are associated with the increased diversity. Streams have more diversity over ponds because of more disturbances. Using 1-, 2-, and 3-patch simulation models, Wu et al. (1993) simulated source–sink interactions, persistence, and resilience of populations. Increased patchiness correlated directly with persistence and resilience. Such simulation experiments are reflected in experiences with natural systems. For example, Walde (1991) experimented with predator and prey mites on apple trees in groups of 1, 4, and 16 trees. The largest population densities, and most persistent populations, were associated with the largest groups of trees. Similarly, using patches of meadows, Robinson et al. (1992) determined that persistence of individuals in patches increased with patch size. A scaling result was also discovered: larger-bodied mammals did best on larger patches; smaller-bodied mammals on smaller patches. Small-mammal distribution represents a source–sink pattern with the smaller patches providing the source of individuals. These small mammals competed directly for resources and were able to coexist simply because of their natural partition of different patch sizes.

Landscape Ecology

As discussed earlier, different natural processes occur at different temporal and spatial scales. Land management professionals typically focus their efforts at temporal resolutions of weeks to decades and spatial resolutions of hectares to thousands of hectares. Landscape ecology is a discipline that shares this focus (Turner 1989). Delcourt and Delcourt (1991) recognize the importance of understanding modern landscapes by reflecting on the changing state of landscapes over the Quaternary Period (past 1.8 million years) and especially the Holocene Epoch (last 10,000 years). Landscape ecology is especially concerned with the human–nature interface at the landscape scale and tends to take a holistic approach to these processes. This approach is reflected in a large number of landscape indices that attempt to capture the essence of the entire landscape in a few numbers. Examples include measures of most common characteristic (dominance), connection of a common characteristic across an area (contagion), and the repeatability of spatial patterns (fractal dimension) (O'Neill et al. 1988). Several dozen such measures have been developed. Many of these change with spatial scale (Turner et al. 1989).

Another set of indices associated with landscape ecology has come out of percolation theory. This theory concerns itself with the patterns of patches on the landscape and, specifically, the probability that patches of certain dimensions and randomness form corridors that span across the landscape. Caswell (1976) developed the notion of "neutral models," simplified computer landscapes that are randomly generated to provide patterns of suitable and unsuitable habitats. For example, assume an animal or plant is constrained to live only in the suitable habitat and has no possibility of even crossing unsuitable areas. A number of questions can be posed to such a system. Gardner et al. (1991) demonstrated that below a landscape coverage of 0.6 (60%), patches are highly fragmented. Their simulations demonstrated "that large differences in species abundance and habitat utilization are produced by small changes in the maximum possible dispersal distance." Turner et al. (1989) used percolation models to evaluate disturbance intensity and frequency on various densities of habitat in neutral maps. Disturbance frequency and intensity had variable impact on neutral-model landscapes. When the landscape was occupied by less than about 50% of the habitat, that habitat was sensitive to frequency, but demonstrated little difference in its response to intensity. Habitats occupying more than 60% of the landscape were less sensitive to frequency, but more sensitive to intensity. O'Neill et al. (1992), through random models, showed that hierarchically structured landscapes (vs. random neutral-model landscapes) had smaller perimeters, were less clumped on sparse landscapes, and were more clumped on dense ones. This permits percolation on a broader range of conditions.

Clearly, a modern ecological modeling environment must provide the modeler with opportunities to develop spatially explicit systems. The patchwork found in the matrix of landscapes is an essential component to the processes that determine population densities. Patches are probably important at every spatial scale, although most research has focused at a scale in which individuals or metapopulations are resolved.

3.2 Ecological Simulation Software

Ecological principles, theories, and experimental data have resulted in a large number of computer-based models and modeling environments. This section briefly explores the extent of this development effort.

Animal Simulations

Computer-based simulations of animals can be grouped into three main categories: (1) individual-based simulations for the study of behavior, genetics, and evolution, (2) theoretical metapopulation models, and (3) population-based models for watershed and region management. Several individual-based models for exploring fish population responses have been developed by DeAngelis et al. (1998) Individual-based modeling has also been used to capture the behavior and energetics of the wood stork on 15-minute time intervals (Fleming et al. 1994). Risenhoover (1997) describes a spatially explicit individual-based deer behavior simulation model called the Deer Management Simulator (DMS).[1] This model was developed for the National Park Service as a tool to help in the management of problems associated with ungulate overpopulation. The DMS integrates land-cover information, initial deer population locations, and deer foraging behavior, and is completed with user menus and interfaces. The software is available for download (Risenhoover 1997). Another example is the Ecosystem Management Model that integrates ARC/INFO (a geographic information system, GIS) with a FORTRAN-based ecosystem landscape model developed to help manage Elk Island National Park in central Alberta, Canada (Buckley et al. 1993). The model consists of an integrated set of submod-

---

[1] DMS—http://www1.nature.nps.gov/dms/dms.htm.

els, including a spatially explicit process-oriented model of vegetation productivity and growth, and an ungulate submodel that takes care of population dynamics, predation, parasitism, and animal condition. Modern computers are allowing us to create simulation models that capture our understandings of nature at lower levels of organization. Spatially explicit individual-based simulation models are now part of a growing trend in the management of natural resources.

Landscape ecologists frequently use population modeling. The demand for software to support population-based simulation modeling is sufficient to support commercial products. Applied Biomathematics supports the RAMAS series of ecological software.[2] Community- and population-based simulation modeling has been supported since the mid-1980s. Metapopulation simulation models allow resource managers to evaluate the importance of interbreeding between two or more populations separated by space. Recently, spatially explicit simulation has been supported in the RAMAS GIS package. Habitat suitability (HS) models are applied to information stored in raster GIS data layers. The resulting suitability maps are automatically analyzed to identify habitat patches that are then fed automatically into the standard RAMAS community- and population-modeling models. This software is founded on metapopulation and patch theory discussed earlier in this chapter (Whigham and Davis 1989, Buckley et al. 1993, Cuddy et al. 1993).

The new discipline of artificial life explores the notions of genetic evolution by simulating competition among different computer programs. These programs are allowed to reproduce, move in a virtual space, mutate, and develop into processes much different from the original programs. Several popular and powerful environments, including Tierra, SWARM, and Poly-World, are currently available to artificial life researchers. The Tierra simulator is a system for studying ecological and evolutionary dynamics (Ray 1994a,b). It is a virtual computer and operating system that allows the execution of machine code that can evolve and compete with other programs. The architecture of the virtual computer can be changed and users are provided a wide range of opportunities for developing software programs that operate within this environment. SWARM is a similar environment for developing and exploring the emergent behavior of simulated life forms (Hiebler 1994, Minar 1995). The goal of SWARM is to have a standardized set of tools for exploring complexity. The concept of a swarm is that sets of similar entities can be grouped together and treated as a unit. This unit itself may be combined with similar units to form hierarchies of swarms. Each entity in the swarm operates independently; behavior is based on the internal and external system states. PolyWorld was developed by Sun computers as a single, powerful artificial life system that "attempts to bring together

---

[2] RAMAS Ecological Software—http://www.ramas.com.

all the principal components of real living systems into a single artificial living system" (Yaeger 1993). It offers a wide range of behaviors (eating, mating, fighting, moving, turning, focusing, and lighting) and provides its simulated life with neural network learning capabilities. Its intended audience is evolutionary ecologists. These and other artificial life programs can be explored on the Internet. The Massachusetts Institute of Technology (MIT) Artificial Intelligence Laboratory[3] and the Santa Fe Institute[4] are excellent starting points (Bodelson and Butler-Villa 1995, Thau 1995).

Landscape Simulations

Simulation of ecological processes at the level of the landscape has also resulted in a significant number of models and modeling approaches. Forest ecologists have been very productive in this area, producing a number of modeling environments. An example is the JABOWA model (Botkin et al. 1972, Botkin 1977). It is an individual-based model that tracks the growth of trees and their effects on their neighbors within a small area (about $10\,m^2$). The loss of large trees within such an area leaves a gap in the forest canopy. More recent versions of gap models simulate a large number of "gaps" that match cells in a raster GIS. One such model is ZELIG, a dynamic simulation environment that divides landscapes into cells divided into gap-scale plots (Urban et al. 1991). The plots are identified with the proportion of total area in different cover types. Another example is LANDIS, a JABOWA/FORET model simultaneously run for each cell in a large raster matrix. LANDIS was developed by Mladenhoff et al. (1993). Individual trees are modeled as part of cells that consider the size, location, type, and state of all member trees. Models in neighboring cells are allowed to dynamically affect each other using this approach.

A large number of modeling approaches based on patch theory (see previous discussion) are represented by the following examples. PatchMod is (1) a spatially explicit age- and size-structured patch demographic model and (2) a multiple species plant population dynamic model. PatchMod was used to model the Jasper Ridge serpentine grassland; gopher mounds provide the primary patch-generating disturbance (Wu 1994). The ARC/INFO GIS and a FORTRAN-based ecosystem landscape model were combined through an ecological modeling interface to address vegetation and ungulate management objectives. The natural system is broken down, for model development purposes, into 12 primary submodels (Buckley et al. 1993).

Numerous estuarine and river basin models have been developed through a cellular modeling approach to assist land managers and politi-

---

[3] MIT Artificial Intelligence Laboratory—http://www.ai.mit.edu/.
[4] Santa Fe Institute—http://www.santafe.edu.

cians in selecting the best long-term land-use decisions (Costanza et al. 1986, 1990, Costanza and Maxwell 1991). More recently, the approach has been used to develop a Patuxent landscape model (Costanza et al. 1993), and an Everglades landscape model (Costanza et al. 1992). The software supporting these models is called the Spatial Modeling Environment (SME) and is currently under development by its author (Maxwell 1995). Another application for cellular-based simulation models is forest fire modeling (Rothermel 1972, Clarke et al. 1993, Kessell 1993, Concalves and Diogo 1994). Such models are typically cell-ased and employ relevant physics-based governing equations and appropriate stochastic functions to capture uncertainty in such things as the lobbing of embers from exploding logs and shifts in swirling winds. In the forest fire model, fire enters a cell where it burns available fuels, expands to neighboring cells based on fire intensity, local slope and elevation, and the current wind, temperature, and humidity conditions.

In this chapter we have taken a quick tour through a number of basic (and competing) ecological theories that individually and in part provide a foundation for watershed simulation modeling. The appropriate scale for a model is one step below the level being investigated. Modeling a watershed requires a focus on the main components of that watershed, including populations of vegetation species and very small animal species, behaviors of large individuals, and the movement of water and air through the system. A selection of individual- and population-based simulation models provided a glimpse into the modeling possibilities that are being used in watershed management to support individual and population modeling. The next chapter similarly looks at the field of hydrology. Chapters 5 and 6 in Part II then further explore watershed modeling opportunities; they are organized by management decisions.

4
Perspectives in Hydrologic Modeling and Simulation

Unlike ecology, the field of hydrology is rich with fundamental and well-accepted classical physics-based equations. These equations are universal—water behaves identically regardless of where on earth it is studied. This is in dramatic contrast to ecological modeling where fundamental system behavior algorithms are not available and the salient processes of the system vary dramatically among locations. There are very few ecological models suitable for landscape and watershed management and they are usually developed to be suitable for a particular location. Typically, the most efficient way to develop ecological models is to begin with simulation modeling environments, within which the local ecological interactions and behavior can be specified. Hydrologic models, on the other hand, are numerous; there are many different models, and user interfaces are well developed. Singh (1995) provides an excellent and thorough review of and introduction to hydrologic simulation modeling.

The maturity of hydrologic simulation modeling is matched with a rich variety of alternative approaches and focuses. There are computer-based simulation models that represent every phase of the water cycle, including climate and weather, storm systems, rainfall, overland water flow, transpiration, overland storm-water flow, groundwater, stream and river flow, and currents and tides in surface-waters in bays and estuaries. Those steps that are associated with watershed management involve three distinct aspects: the movement and flow of water, the associated erosion, and the movement of crop nutrients, pesticides, and herbicides. Models have been developed to simulate all of these processes.

Water movement through a watershed occurs via overland flow, movement through unsaturated and saturated soils, and flow down streams and rivers. Water flow across the ground or down streams and rivers is a function of the amount of water, the slope, and the surface roughness. Simulation models are used to predict water flow rate, depth, and scouring and sediment load potential. They must take into account surface roughness and the related resistance to flow, water momentum, surface slopes, stream and river depths, and soil characteristics. Hydrologic modeling is a very active area of research and development motivated by potential losses to property, loss of life, and threats to health due to nonpoint source pollution.

4.1 Simulation Models

A number of representative models, divided into the categories listed below, are briefly reviewed. This review provides a sense of the scope and depth of hydrologic simulation models developed to understand and predict hydrologic behavior of and within watersheds.

- Field-scale hydrologic and soil erosion models
- Watershed-scale hydrologic and soil erosion models
- Groundwater models
- Field-scale water quality models
- Watershed-scale water quality models

Field-Scale Hydrologic and Soil Erosion Models

Field-scale models treat entire fields as single, discrete, and homogeneous entities. These models are typically simple enough to state, but difficult to parameterize, which can often result in a handbook with a simple equation followed by many pages of look-up parameters. When a field does contain a single dominant soil type, has a constant slope and aspect, and a single management history, these models are quick and efficient. Field-scale hydrology models have been developed to predict anticipated farm field erosion based on weather, climate, field conditions, crop, soil type or qualities, and topography. While these models have been developed to assist in the management of farm and grazing lands, this management occurs in the context of watersheds and the models can be useful beyond the farm. Before computers could be applied, it was necessary to develop and adopt simple equations that could be used by farmers and land managers. One such model is the "rational method." It calculates the peak runoff rate as follows:

$$Q = 0.002CiA \qquad (4.1)$$

where Q = peak runoff rate
      C = a dimensionless runoff coefficient
      i = rainfall intensity
      A = the watershed area in hectares (ha)

This equation has many assumptions including steady state watershed outlet flow due to constant rainfall in time and space over the watershed. It also assumes no infiltration. Finally, the equation does not predict the time of peak flow or any other part of a hydrograph's structure.

Other simply structured equations developed in the mid–twentieth century are the Universal Soil-Loss Equation (USLE) and the updated Revised Universal Soil-Loss Equation (RUSLE) (Wischmeier and Smith 1978). For both, the form of the equation for predicting field soil loss is identical:

$$A = RK(LS)CP \tag{4.2}$$

where A = predicted average annual soil loss
     R = index for local rain dislodgment of soil and movement of soil in runoff
     K = factor for soil erodability
     LS = factor for slope and length of slope
     C = factor for crop cover
     P = factor for conservation practices

A number of indices reflecting rainfall erosivity, soil erodability, slope length, steepness, cover, and conservation practices are identified for the location of interest and multiplied together to estimate the average annual soil loss. Indices have been developed through many years of experimentation and trials. By using look-up tables, one can readily apply this simple model to an area to calculate average annual soil loss. Some limitations of the model were overcome with the introduction of the RUSLE, which includes modern Windows-based user interfaces that allow for automatic table look-up based on user specification of slope, soil types, crop cover, and location. The RUSLE and USLE parameters were developed over many decades of measurements followed by statistical analyses. Limitations include the need for erosion studies when different soil types and crop covers are encountered. Also, while sheet and rill erosion is considered, erosion associated with gullies (created when rills converge) is not estimated. These limitations are addressed through the development of process-based models and through the application of geographic information systems (GISs).

The USLE and RUSLE presume that the area under consideration is relatively homogeneous with a constant slope and aspect and containing a single soil type, land cover, and conservation practice. Moving to larger parcels eventually ensures that the terrain is complex over space and over time. If a complex landscape is divided into smaller parcels and those parcels are hydrologically connected, then the USLE/RUSLE analyses can be completed for each parcel. Systems such as the Areal Nonpoint Source Watershed Environmental Response Simulation (ANSWERS) model (Beasley and Huggins 1982) and Agricultural Non-Point Source (AGNPS)[1] pollution model (Young, Onstad et al. 1989) provide this approach. These

---

[1] AGNPS—http://www.cee.odu.edu/cee/model/agnps_desc.html.

models still use the various experimentally derived USLE and RUSLE indices and factors. A next step is to develop physics-based process models that can be applied to any area where the physical properties and components of the soil are known. Spatially explicit process-oriented erosion simulation models include Cascade-2D (CASC2D) (Saghafian 1993), the Water Erosion Prediction Project (WEPP)[2] model, and the Simulation of Watershed Erosion (SIMWE) (Mitasova et al. 1998). These are process-based distributed parameter models that run in conjunction with digital map inputs. The inputs include topographic information, such as slope and elevation, soil qualities, and crop cover; weather information, including synthetic or recorded storms; and field treatment schedules. The physical processes involved when rain dislodges soil and when sheet, rill, and stream flow moves dislodged particles are modeled. Pure physical process-based models can be run to develop the USLE and RUSLE model parameters on areas that have not been studied. Information about the soil structure, such as percentages of sand, clay, loam, and organic matter, is typically required. These models are computationally intensive and have become useful only recently with the cheap availability of fast computers.

As we moved through this list from the experimental and statistically based USLE and RUSLE to the process-based models, the computational requirements increased dramatically. Adoption of spatially explicit modeling becomes important for complex terrains in which the topography, cover, and/or treatment varies. Adoption of process-based modeling allows us to simulate complex terrains without requiring that indicies and factors be preestablished. Process-based modeling requires the application of powerful (but now inexpensive) computers, while the USLE and RUSLE approaches can be accomplished with a handbook, pencil, and paper. The more complex models will continue to become increasingly cost effective as the models and model input data become easier to acquire and use.

Watershed-Scale Hydrologic and Soil Erosion Models

Within a watershed there can be hundreds of separate fields. If those fields can be appropriately modeled with field-scale models, it should be possible to describe the watershed processes by combining all of the field models. While it is possible to model larger watersheds using field-scale simulation models, the data requirements can become overwhelming. Complex terrains are very difficult to model with field-scale models and combining a large number of fields in such a terrain is not likely to yield useful information about the watershed as a whole. It has traditionally been popular to model watersheds as whole entities. These watershed-level models are

---

[2] WEPP—http://topsoil.nserl.purdue.edu/weppmain/wepp.html.

more frequently built up from lumped-parameter models of subwatershed components. Two such models are TR-20, developed by the Soil Conservation Service, and HEC-1, developed by the U.S. Army Corps of Engineers' Hydrologic Engineering Center. The hydrologic response of subwatershed components (overland flow areas and stream/river segments) are defined and connected to allow for full watershed hydrologic responses. These programs were developed at a time when input data were provided through punch cards and, today, the programs require input provided via card images in computer files. A modern user interface has been developed for the Watershed Modeling System (WMS)[3] that automates the development of the card image files through automatic interactions of watershed information stored in GIS data files. Lumped-parameter models require that each watershed subcomponent be characterized with a set of numbers representing its general or cumulative nature. Systems such as the WMS query raster and vector GIS data to automatically generate the lumped parameters for each of the subwatersheds and associated streams and rivers. Users can still be responsible for identifying detailed stream/river cross-section information. Recently, the Hydrologic Engineering Center released a new product intended to supercede HEC-1 called HEC-HMS (Hydrologic Modeling System).[4] A modern graphical user interface (GUI) and standard database system has been fully integrated into the modeling system.

Groundwater Models

Like watershed-scale models, groundwater models were originally developed for computers that took input through computer cards. Many of these historic models are finding new life in integrated systems that run on modern desktop and workstation computers. An excellent example of such a system is the U.S. Army Corps of Engineers' Groundwater Modeling System (GMS).[5] A number of models have been combined and incorporated into this system (Owen et al. 1996). MODFLOW[6] partitions a subsurface area into discrete three-dimensional (3D) chunks that are each defined by location and soil characteristics. Water flows are routed through this space. MODPATH[7] routes particles through this space. SEEP2D assists in the modeling of water flow under and through dams and levees. A number of pollutant movement and tracking models (see below) are associated with the water movement model.

---

[3] WMS—http://ripple.wes.army.mil/software.
[4] HEC-HMS—http://www.hec.usace.army.mil.
[5] GES—http://www.hec.usace.army.mil.
[6] MODFLOW—http://water.usgs.gov/software/modflow-88.html.
[7] MODPATH—http://water.usgs.gov/software/modpath.html.

Field-Scale Water Quality Models

Hydrologic models route water over and through the ground. Water movement facilitates the transport of various chemicals—some of which influence water quality. Many water quality models have been developed for field-scale settings that are concerned with the movement of nitrogen, phosphorus, potassium, and various organic herbicides and pesticides. The Root Zone Water Quality Model (RZWQM)[8] simulates the movement of water and associated chemicals in the vertical direction as part of an integrated crop growth modeling system. The Groundwater Loading Effects of Agricultural Management Systems (GLEAMS), the Erosion/Productivity Impact Calculator (EPIC), and the Chemicals, Runoff, and Erosion from Agricultural Management Systems (CREAMS) are examples of coupled two-dimensional (2D) field-level simulation models for predicting chemical transport. Discussions about and availability of these and other models can be found at a Natural Resources Conservation Service Internet site.[9]

Watershed-Scale Water Quality Models

Movement of water, soils, and chemicals is, of course, also modeled for larger watershed systems. Different systems predict water quality in urban and rural watersheds. In the GMS (Owen et al. 1996), a number of models (MT3D, RT3D, and FEMWATER) simulate the movement of contaminates through the groundwater. The Water Quality Analysis Simulation Program (WASP), a DOS program developed by the U.S. Environmental Protection Agency (EPA), combines a number of other models that simulate hydrodynamics, unsteady flow in one-dimensional rivers, unsteady 3D flow in lakes and estuaries, conventional pollution (involving dissolved oxygen, biochemical oxygen demand, nutrients, and eutrophication), and toxic pollution (involving organic chemicals, metals, and sediment). The EPA's Storm Water Management Model (SWMM)[10] is similarly DOS-based with a long development history that includes a number of modern Windows interfaces. It models single-event and continuous watershed water quality simulation primarily, but not exclusively, for urban watersheds.

The Soil and Water Assessment Tool (SWAT)[11] is a public domain product under active development at the Agricultural Research Service's Grassland, Soil, and Water Research Laboratory (Temple, TX). SWAT employs a modern Windows interface to an integrated system of models

---

[8] RZWQM—http://www.gpsr.colostate.edu/GPSR/products/rzwqm.htm.
[9] Water, Field Scale and Watershed Scale Computer Models, Field and/or Point Assessment Tools, and Tools Under Development—http://www.wcc.nrcs.usda.gov/water/quality/common/h2oqual.html.
[10] SWMM—http://www.epa.gov/SWMM_WINDOWS.
[11] SWAT—http://www.brc.tamus.edu/swat.

and GISs that routes water and chemicals through surface flow, groundwater flow, and stream/river flow, and can be applied to watershed basins of several thousand square miles.

The Hydrologic Simulation Program-FORTRAN (HSPF)[12] is a comprehensive modeling set that simulates the movement of pollutants (conventional and toxic) through land/soil runoff processes linked directly to in-stream chemical and hydraulic processes. Model output includes flow rates, chemical concentrations, and sediment loads.

All of the models listed is far require an operator trained in hydrologic modeling and comfortable building input files in a DOS environment. Although the modeling equations are all captured in software, the parameterization of the model and data collection for a particular location can be arduous. This makes virtually all of the models inaccessible to watershed planning groups except through the expertise of water quality and hydrologic engineers. The EPA has worked very hard to create a watershed water quality modeling environment that is accessible to more people. The system is called Better Assessment Science Integrating Point and Nonpoint Sources (BASINS).[13] This system is based directly on the commercial GIS, ArcView, and its native user interface and programming language, AVENUE. Through Internet connection or CD-ROM, users access not only the set of programs, but preformatted GIS data required to run the model. Virtually any watershed in the United States can be modeled using readily available and preformatted data. Users are also provided tools and instructions for updating the data to reflect local policy and construction changes. BASINS offers a nonpoint source model (NPSM), which is a user interface combined with HSPF (see above). It uses QUAL2E for steady state water quality and eutrophication modeling and TOXIROUTE for simple dilution/decay of pollutants for screening purposes. The ArcView-based user interface makes the model accessible by individuals familiar with the ArcView GIS environment as well as hydrologists.

4.2 The Role of Geographic Information Systems

Nonpoint source pollution simulations have involved the linkage of various simulation packages with GISs. For example, ANSWERS (Areal Nonpoint Source Watershed Environment Response) has been linked with GISs (Rewerts and Engel 1991, Krummel et al. 1993). Hay et al. (1993) have integrated GUIs, statistical analysis packages, and GISs. Srinivasan (1992) combined a GIS with the SWAT simulation software.[14] Several different water

---

[12] HSPF—ftp://ftp.epa.gov/epa_ceam/wwwhtml/hspf.htm.
[13] BASINS—http://www.epa.gov/OST/BASINS.
[14] SWAT GRASS—http://www.baylor.edu/~Bruce_Byars/swatgrassman.html.

quality (WQ) models from the Agricultural Research Service have been captured within the GRASS (Geographical Resources Analysis Support System) GIS (Cronshey et al. 1993). DePinto et al. (1993) linked the geographically based WAMS (Watershed Analysis and Modeling System) model with the ARC/INFO GIS to simulate watershed loading, groundwater contaminant transport, and dissolved oxygen in rivers.

In many cases, water flow simulation software has been developed for the purposes of a single project. For example, D'Agnese et al. (1993) established a rather complicated mix of several software packages to develop a series of groundwater simulations for an area near Death Valley, CA. Frederickson et al. (1994) integrated GIS, HEC-1, HEC-2, and a GUI to develop a flood impact prediction prototype for the U.S. Army Corps of Engineers' Omaha District.

All of the above examples read (and sometimes write back to) GIS databases. The GIS software is not used to accomplish the simulations. Instead of using the GIS only as a data source, new approaches process raster GIS data in their native cellular format. For example, a general-purpose finite-element approach to watershed overland flow simulation called r.fea has been designed and developed within the GRASS GIS (Gaur and Vieux 1992, Vieux and Westervelt 1992, Vieux et al. 1993). A finite-difference approach (CASC2D in GRASS) has been developed by Saghafian (1993) and others.

A growing number of hydrologic models at watershed and field scales have been tightly integrated with GISs. Those linked with the public domain GRASS GIS include AGNPS, ANSWERS, CASC2D, GLEAMS, SWAT, RZWQM, WEPP, MODFLOW, SIMWE, and others not covered in this text. Many are also being linked with the ArcView GIS from ESRI[15] including WEPP, HSPF, and the watershed manager–oriented BASINS. These models are but a few examples of the simulation modeling that has been formally developed to route water and associated sediments and soil through a watershed. Hydrologists have many competing approaches to effectively and efficiently move water in simulation space, and have done so for overland flow, stream and river flows, and groundwater flow. Because these models have been developed to advance the understanding of watershed processes, they are often not ready, or appropriate, for directly advising watershed managers.

This chapter introduced watershed simulation modeling from the perspective of the responsible scientific disciplines (and subdisciplines). The next section looks at simulation modeling from the standpoint of the needs of watershed managers. The ideal system would be a fully linked multidisciplinary, multiscale simulation model representing not only the ecology and hydrology (covered in this chapter), but also the associated economic

---

[15] ESRI—http://www.esri.com.

and social systems. Managers would like the equivalent of a map that has a scale of "12 inches to the foot." While interdisciplinary simulation modeling will dramatically develop in the coming decades, it will still be important to choose from the available models based on the precise management questions being asked within the constraints of the current resources. Part II provides guidance that will help watershed managers now and well into the future with these choices.

Part II
Choosing Models and Modeling Environments

In Part II, we pragmatically explore approaches to addressing different land and watershed management concerns using dynamic simulation modeling approaches. These approaches range from the application of already developed and well-supported software to the development of locally specific simulation models. The questions are grouped into two main categories: those addressed with single-discipline simulation models (Chapter 5), and those requiring multidiscipline scientific models (Chapter 6). This basic taxonomy moves us from basic science models through interdisciplinary science models and into models captured by management-oriented decision support systems (DSSs). This overview shows both the breadth of capabilities and the lack of good general-purpose multidisciplinary decision supports systems. In Chapters 7 and 8 we consider approaches to the design and development of locally explicit simulation modeling—a step often required when environmental issues are involved. This is followed by considering software assistance in the evaluation of alternatives through trade-off analyses that capture the values of the decisionmakers (Chapter 9) Finally, a decision process is presented to help managers select software models and modeling environments for their applications (Chapter 10).

5
Questions Addressed with Single-Discipline Simulation Models

In the previous chapters, traditions in hydrology and ecology were introduced and associated with sample simulation models developed out of those traditions. This chapter takes a more management-centric view and approaches models from sample management decisions and questions. Each section is introduced with a small number of representative questions and decisions, and then identifies how simulation modeling can be effective in addressing those management needs. You are encouraged to extend the issues posed with issues that might currently concern you and then develop solutions that might be afforded by the models presented in this book.

5.1 Questions Addressed by Surface Water Erosion and Pollution Models

Sample management questions/concerns:

- How are my land management patterns and practices related to stream hydrographs in severe storms?
- What water quality is expected downstream as a result of my land management practices and patterns?
- If we rezone to allow a particular land management practice, what will be the consequences to surface water, groundwater, and downstream water quality?
- How can I plant grasses in a field to get the maximum decrease in expected sediment downstream?

Local watershed management groups can come into existence as a result of these types of concerns. Individuals and businesses located near a stream or river are concerned with the flooding potentials that change as a result of upstream land management. Those who make use of the water in those streams for drinking or recreation are concerned with upstream deposits of chemicals and nutrients that might degrade the water quality. The types of questions posed here are readily addressed by existing water simulation models. Some of those models are identified below, but they do not exhaust the growing list of possibilities.

Erosion associated with agricultural practices has been responsible for the development of a large number of computer software–based modeling and simulation products. They seek to predict erosion based on current or expected land management (cropping) practices. A number of software products have been developed to evaluate alternative land management practices at the level of a farm field. The Water Erosion Prediction Project (WEPP)[1] runs under DOS and Windows 95, 98, and NT environments. Its spatially explicit input requirements include topography, crop management, storm characteristics, and soils. WEPP generates erosion and deposition predictions associated with sheet, rill, and water channel processes at the level of a field. Simulation of Water Erosion (SIMWE)[2] uses similar inputs and computes the spatial distribution of flow, erosion, and sedimentation rates during a steady rain. It runs in a number of computer hardware and operating system environments and is useful for optimizing land-use patterns for minimizing erosion and deposition problems. Chemicals, Runoff, and Erosion from Agricultural Management Systems (CREAMS) (Kinsel) is also a field-level system, but it is not spatially explicit. Nonpoint source pollution prediction for agriculture watersheds is the focus of the Agricultural Non-Point Source (AGNPS)[3] pollution model (Young et al. 1989). It is a spatially explicit simulation model that requires a large number of inputs, including topology derivatives, soil characteristics, land cover, watershed channels and impoundments, fertilization information, chemical factors, and storm characteristics. AGNPS outputs watershed information on erosion and deposition, chemical concentrations over time, sediment mass loads, and concentration of materials. The watershed manager can compare alternative land management scenarios with respect to the output information.

The U.S. Environmental Protection Agency's (EPA's) premier software for evaluating pollution in drainage water is Better Assessment Science Integrating Point and Nonpoint Sources (BASINS)[4]. BASINS combines a commercial GIS (ArcView from ESRI) with national environmental databases, watershed assessment extensions, and characterization reports to ArcView, and models for stream water quality and nonpoint source simulation. Data for driving BASINS can be downloaded from the main BASINS World Wide Web (WWW) site. The data include political boundaries, watershed delineation, digital elevation maps, soils, land cover, ecoregions, water quality and gauging, wildlife, minerals, and others. BASINS can

---

[1] WEPP—http://topsoil.nserl.purdue.edu/weppmain/wepp.html.
[2] SIMWE—http://owww.cecer.army.mil/td/conservation/find/factsheet.cfm?id=323.
[3] AGNPS—http://www.cee.odu.edu/cee/model/agnps_desc.html.
[4] BASINS—http://www.epa.gov/OST/BASINS/.

be downloaded at no cost over the Internet and can also be installed from CD-ROM.

5.2 River Management

Sample management question/concern:

- Our river is a source of wealth, inspiration, and recreation. Plans exist to construct levies to protect property, dam the river to control downriver flooding and provide irrigation, and establish new construction in the floodplains. How will the risks of damage by flooding change under these scenarios?

This sample concern is related to those in the previous section, but it provides more of a focus on the behavior of a stream or river in response to efforts to directly manage that stream or river. The management of navigable waters in the United States at this level has been the responsibility of the U.S. Army Corps of Engineers, and has resulted in a rich history of software development. Emerging new capabilities provide user interfaces that are adapted to a much larger user base than earlier capabilities, but the underlying mathematics remains solid and consistent. The history of these models has been long enough to establish them with some level of legal weight and authority.

The movement of water in streams and rivers is very well understood; it has been possible to develop straightforward models that can be used to predict water flow depths, velocities, and flooding extents. A number of models developed by the U.S. Army Corps of Engineers' Hydrologic Engineering Center (HEC) have been used for the past several decades for assisting with river management. The Hydrologic Engineering Center has merged a number of their traditional capabilities together behind a graphical user interface (GUI) into the Hydrologic Modeling System (HEC-HMS).[5] This system runs under the Windows 95, 98, and NT environments as well as X-windows–based UNIX systems. It is a fully integrated overland water flow and stream routing system that can accept real storm events defined through Doppler radar. The land manager provides model inputs, including soil characteristics, digital elevation, land-use patterns, and other basin data. Outputs include storm-water runoff velocities, depths, and volumes. The fundamental simulation unit is a watershed—defined by the modeler—and a hydrologic routing model that represents the course and cross-sectional characteristics of streams and rivers in the study area. The Hydrologic Engineering Center also offers a River Analysis System (HEC-RAS) that supports the calculation of steady state flow water surface profiles that take into account the effects of various structures, such as bridges,

---

[5] HEC-HMS, HEC-RAS, and HEC-FDA—http://www.hec.usace.army.mil.

culverts, weirs, and buildings. A GUI supports model design and development, model execution, and the graphical and tabular display of outputs. Future versions are scheduled to support unsteady flow and sediment transport. A third major HEC program is called Flood Damage Analysis (HEC-FDA), which facilitates a flood-impact reduction planning process. It provides an integrated hydrologic engineering and economic analysis platform intended for use by river managers and engineers. All three of these programs, supporting documentation, and example applications are available without U.S. Army Corps of Engineers support at no cost through Internet access.

The Coastal and Hydraulics Laboratory at the Waterways Experiment Station, a U.S. Army Corps of Engineers Research and Development facility, has also developed a suite of watershed and water management programs. These are the Groundwater Modeling System (GMS), Surfacewater Modeling System (SMS), and Watershed Modeling System (WMS).[6] These three systems share a similar look and feel, but each focuses on a different part of the water system. All allow for the rapid construction of models using a variety of graphical interface–driven tools that accommodate GIS data layers imported from various popular GIS software packages. The GMS provides a collection of tools to build subsurface models and establish groundwater flow boundary conditions, and then allows the user to use any of a variety of groundwater simulation models. Powerful graphical tools allow for the visualization of the three-dimensional (3D) data space inputs and outputs. The GIS data layers, borehole information, stream data, and other local data sources can be accommodated in the construction of the basic subsurface model through which water flow is simulated. Two- and three-dimensional mesh, grid, and scatterpoint modules are used to interpolate known data to form surfaces and solids. The SMS provides similar model construction tools and then focuses a variety of 2D shallow open-water models to represent the behavior of estuaries, rivers, bays, and wetlands. The WMS targets the overland flow and infiltration of water within watersheds. Like the other two systems, WMS allows the user to characterize a watershed using GUIs, input of GIS data and satellite imagery, and then simulation of the water flow associated with storms. A number of different numerical methods are available. One of the models behind WMS is CASC2D, a 2D overland water flow simulation model. This model divides the land surface into a grid of patches, each of which is characterized by its elevation, land cover, soil properties, soil saturation, and surface roughness. Spatially explicit rainfall events that vary over time (e.g., from Doppler radar analyses) can be input into the model. Using fundamental hydrologic equations (the Green and Ampt equations for infiltration and diffusive wave approximations of the St. Venant equations), a finite-difference grid-

---

[6] GMS, SMS, and WMS—http://chl.wes.army.mil/software/.

based scheme routes water over the landscape and into stream networks. Equations supporting the flow of water in channel systems round out the CASC2D model. This model has also been incorporated into other GIS and landscape simulation environments.

5.3 Stream Management

Sample management question/concern:

* A stream runs through our city. During relatively severe storms the stream overflows and floods a few blocks in an important commercial area. What alternatives does the city have to minimize flooding?

This concern is similar to that in the previous section, but at a much smaller and local scale. As noted above, the U.S. Army Corps of Engineers is responsible for managing navigable waters, but the management of local streams remains in the hands of local officials. Fortunately, many of the models developed by the U.S. Army Corps of Engineers are available and applicable at these smaller scales. Let us assume in this situation that the stream flooding is based primarily on the management of land associated with the city. As cities develop, the amount of permeable land decreases. Originally vegetated soils are replaced with buildings (impermeable rooftops), parking areas, and streets. Where rain was originally absorbed by the soil and then, through groundwater flow, seeped into local streams, that rain is now moved rapidly and efficiently into the local stream through surface flows. Also, original streams and brooks are often transformed into straight and deep ditches that efficiently drain storm waters. How do we quantify these effects with respect to alternative land management plans?

The EPA's Storm Water Management Model (SWMM)[7] simulates the movement of water and pollutants from a rainfall event through an urban drainage system and into streams and rivers. SWMM can accommodate a single event and continuous rainfall. A Windows 95, 98, and NT version of SWMM (PCSWMM)[8] simplifies the development, visualization, and operation of models.

The EPA's Water Quality Analysis Simulation Program (WASP)[9] combines the following two subsystems: the Toxic Chemical Model (TOXI) and the Eutrophication Model (EUTRO). TOXI predicts chemical concentrations in waters and their beds, while EUTRO predicts phytoplankton, carbon, chlorophyll, ammonia, nitrate, nitrogen, orthophosphate, and

---

[7] SWMM—http://www.epa.gov/ostwater/SWMM_WINDOWS/, http://www.chi.on.ca/swmm.html/.
[8] PCSWMM—http://www.chi.on.ca/pcswmm.html.
[9] WASP—ftp://ftp.epa.gov/epa_ceam/wwwhtml/wasp.htm/.

oxygen concentrations. A commercial version of this software (WASP5[10]) along with a GIS-based software environment for building WASP input data is available from Colorado State University, WASP Builder.[11]

5.4 Plant Community Succession

Sample management questions/concerns:

- We want to restore stable native vegetation to a (roadside?). What species combinations (plant communities) should we expect to exist over time?
- A section of land will be taken out of agriculture. What should be planted there to encourage the rapid establishment of a native community?
- Tracked-wheeled vehicles continually tear up erosion-preventing vegetation. What can I plant to help minimize erosion without an accompanying loss of training time?

These concerns focus clearly on the ecology of a portion of a local watershed. These questions seek to have a portion of land reinitialized with a mix of vegetation that will stabilize the land with minimal on-going maintenance and treatment. One approach is to consult texts on native local ecology, which are frequently available, and to then consult with local nurseries that specialize in native species. However, new computer-based software environments are now available for consultation.

VegSpec[12] is a new WWW-based decision support system (DSS) designed to address these kinds of management concerns. It does not rely on simulation modeling (at least directly). By identifying soil types, simple landscape characteristics, nearest weather station, and management objectives, VegSpec can be used to recommend appropriate plants to consider along with site preparation requirements and seeding rates and instructions. It is most useful for relatively high-intensity landscape management.

A representative emerging simulation-based capability designed to support up-front land treatment and planting decisions is a system called Ecological Dynamics Simulation (EDYS) (Price et al. 1997, McLendon et al. 1998). The goal of EDYS is to predict the anticipated succession of plant communities on a site based on initial starting conditions. With this tool, landscape/watershed managers can better evaluate the long-term consequences of an up-front land management investment. Like VegSpec, EDYS requires soil characteristics, weather and climate information, and an initial population of plants as inputs. The program's outputs are predicted densities of plants over the next several dozen years. EDYS accomplishes this

---

[10] WASP Builder—http://www.chi.on.ca/wasp.html.
[11] http://www.colostate.edu/Depts/IDS/builder/builder.htm.
[12] VegSpec—http://plants.usda.gov/.

prediction by allowing the various plant species to compete with one another for light, water, and nutrient resources. Process models move water from the atmosphere to the ground, through a number of soil layers. Plants are represented with different root and aboveground vegetative growth patterns depending on the species and its growth characteristics under different conditions. EDYS is an excellent example of how knowledge derived from basic research efforts in hydrology, plant physiology, agronomy, and weather can be integrated to create a powerful simulation model to support land management.

Habitat Management

Sample management questions/concerns:

- We are required to protect a local threatened or endangered species. How will our alternative land management plans affect this species?
- We have funds to purchase land management rights from farmers in a watershed. How will species biodiversity be affected by the spatial arrangement of alternative buy-out plans?

These questions are concerned with the management of native wildlife—especially those species that must be protected by law. Hydrologic concerns typically model the behavior of storms and rivers over the course of hours and days. The vegetation growth patterns are concerned with the establishment of native populations over the course of a few years. Management of watersheds to address the needs of threatened or endangered species requires a longer-term evaluation. Also, as we move from hydrology, through establishing vegetation, and into habitat management, we find fewer completed and ready-to-use models. Nevertheless, some starting points are available.

Animal and plant growth and reproduction success are based on environmental context. The physical size of that context is based on the size and movement of the organism. Small plants send out roots, horizontally and vertically, that interact with the available nutrients and moisture, and compete with other plants that are also extracting the resources in the soil. Similarly, the plant competes for sunlight with other plants. Plants interact with their immediate surroundings, and any habitat analysis must be accomplished at this scale. Animals, being mobile, interact with broader components of the landscape. In general, the larger the animal, the larger the area of interaction. Each animal and plant evolved to succeed under particular combinations of environmental conditions within their particular range of interaction. The natural history of a species provides knowledge about the habitat requirements and, for many species, these have been captured in Habitat Evaluation Procedures (HEP). One of the HEP approaches is the application of software that mathematically evaluates a set of environment measures to compute a Habitat Suitability Index (HSI) for various species.

Good examples are the HSI Software Programs developed by the U.S. Geological Survey (USGS) at their Midcontinent Ecological Science Center.[13] These are DOS-based programs that provide habitat suitability indices for several dozen different animals based on input of habitat description information. Spatially explicit GIS-based HSI models have also been developed. The California Wildlife Habitat Relationships Program[14] developed a series of 28 such models based on the ARC/INFO Arc Macro Language (AML) from ESRI.[15] In some cases, locally specific DSSs have built HSI models into the decision support processes. An example is the DSS developed to support management of the Lonetree Wildlife Management in central North Dakota. This system is called the Integrated River Basin Environmental Management (IREM).[16] HSI models for five game species help managers evaluate the impacts of alternative management plans on their populations.

5.5 Urban Growth

Sample management questions/concerns:

- Where will growth occur around my town or city over the next several decades?
- How will land management policies and zoning ordinances affect growth?
- How will a transfer of development rights affect growth?

Planning and management are future-focused activities, and that focus often involves the development of land and watersheds for human settlement. Travel costs encourage businesses, services, and residential areas to be located in relatively close proximity. Established urban areas tend to grow rather than having new urban areas created. However, the detailed patterns of urban growth are a function of the current urban patterns and the spatial arrangement of the natural topography. Predicting the growth patterns based on alternative management scenarios is necessary for urban and watershed planners to choose among the management options.

Urban planners and civil engineers have developed location–allocation models (Ghosh and Rushton 1987) and traffic simulation models (Oppenheim and Oppenheim 1995, Ran and Boyce 1996) to help understand and predict land-use and traffic patterns. Location–allocation models concern themselves with the decision processes that individuals and businesses work

---

[13] HSI Software Programs—http://www.mesc.usgs.gov/hsi/hsi.html.
[14] GIS HSI Models, the California Wildlife Habitat Relationships Program—http://www.dfg.ca.gov/wmd/cwhr/gismodl.html.
[15] ESRI—http://www.esri.com.
[16] http://www.colostate.edu/Depts/IDS/irem.html/.

through to decide where to locate. Such decisions rely heavily on the time required to physically travel between various combinations of locations, which is a function of the geographical layout of the road infrastructure and the traffic load. Traffic simulation models estimate loads on the road network based on trip requirements among the various geographical locations.

Our focus is at a scale of a region, a large watershed, or perhaps a county. Instead of analyzing the flow of traffic throughout a day, we want to focus instead on the growth of urban areas over the course of many decades. Urban growth is driven by a number of factors:

- Strength of the local economy
- Cost of transportation
- Available infrastructure
- Topography
- Attractiveness of the area
- Laws and regulations

Our lives depend on our ability to work to generate an income. The strength of the local economy is responsible for the maintenance of jobs, and, with a strong economy, the net migration will favor urban growth. Low transportation costs (in time and money) encourage growth and expansion. Also, as every developer and city council member is aware, the existence of water, power, sewer, and communication lines is a heavy attractor for development. Construction is least expensive on flatter ground. High slopes deter expansion and, to a more limited extent, so do soils that are not suitable for supporting large buildings. Because it is people that provide the growth pressures, the attractiveness of an area to people is a key factor for development. During the middle part of the twentieth century, Southern California attracted people, despite its distance from other urban centers, because of its weather and proximity to the ocean. Finally, laws and regulations both limit and encourage growth. City and county zoning regulations are used to direct growth in predefined directions. Parks, forests, wildlife refuges, and other similar land uses are powerful levers for locking in future development pattern options.

A recent example of an urban growth model is the Clarke Urban Growth Model[17] (Kirtland et al. 1994). This model was used to predict the urban growth patterns of the San Francisco Bay area and the Baltimore/ Washington, DC metropolitan area. The model supports the following components of growth:

- Spontaneous neighborhood growth
  This is growth at the immediate edge of an urban area prompted by proximity to the urban area and mediated by the ability to build, which is a function of the slope.

---

[17] Clark Urban Growth Model—http://geo.arc.nasa.gov/usgs/clarke/hilt.html, http://geo.arc.nasa.gov/umap.

- Diffusive growth and spread of a new growth center
  This growth creates a new urban area that does not neighbor on an existing urban area. Again, higher-sloped areas are less likely to spontaneously generate a new urban setting.
- Organic growth
  This is growth into areas that are nearly surrounded by urban settings. In this model, cells surrounded by at least three urbanized neighbors are candidates. And, again, slope is a mediating factor.
- Road-influenced growth
  Roads "transmit" urban growth pressures from urbanized areas to areas not in direct contact with the urban area.
- Self-modification
  To simulate the effect of good times following good and bad times following bad, the coefficients associated with growth are increased even higher during high growth and lower during low growth, but mediated to check completely uncontrolled growth and to check complete population crashes.

The Clarke model provides a rich environment for exploring theories of urban growth and for testing the implications of land management and transportation decisions on future settlement patterns.

Urban growth is associated with traffic flow patterns and road and highway construction projects designed to meet current and projected demands. Road and highway construction takes place at smaller spatial scales and over shorter periods of time than multidecade urban growth, but those construction projects are significantly important in defining the growth patterns. To help manage local traffic, cities and regions turn to traffic simulation modeling. These models help planners understand the impacts of alternative traffic management plans on congestion incurred during peak traffic flow times. Simulation time steps can be on the order of seconds to minutes—far different from the week to decade time steps associated with some watershed simulation modeling.

This chapter posed a number of realistic management questions and objectives, and then identified existing simulation models currently available that could be used to address the concerns. Each of these models tends to be centered within a particular discipline. Not all possible management concerns were posed and not all possible approaches for addressing them were presented. Management decisions that involve a single discipline are likely associated with already developed simulation models. Watershed managers should therefore be encouraged to seek out simulation models that might be appropriate for their particular management challenges. Few multidisciplinary simulation models are currently available to support more complex management decisions.

6
Questions Addressed with Multidiscipline Simulation Models

Often, landscape management decisions are associated with competing multiple objectives. Managers do not have the luxury of focusing on a single objective; the best science from a particular academic tradition cannot therefore be tapped for an optimal solution. Consider a situation where flooding concerns must be balanced against ecological goals. Hydrologic models might result in the recommendation that a small dam will help decrease downstream floods, increase irrigation (and therefore economic) opportunities, and provide desired recreational opportunities. Environmental models might indicate that instead of putting a small dam on the stream, a remeandering of the stream and demolition of buildings in the natural floodplain will reap ecological and other economic benefits. It is possible that recommendations from each model will be different if they are linked to more completely represent the full watershed system in a single model. This chapter, like the previous one, poses a small set of example issues intended to resonate with watershed managers. Again, this set is not exhaustive or meant to cover all of the important ground. The reader is encouraged to reflect on the role that simulation modeling might play in their current watershed management challenges.

Sample management questions/concerns:

- We plan to continue using a landscape for military tracked-vehicle training. We have developed alternative land management scenarios that vary training intensity in time and space. Along with our training requirements and objectives, we want to increase biodiversity, maintain erosion-resistant training areas by encouraging native plant growth, and minimize chemical runoff.
- We recognize that urban growth patterns develop in response to economic, social, spiritual, topographical, and natural resource patterns. Our local town, city, and county planning commissions have developed a land-use plan desired 30 years from now. What can be done to ensure that growth progresses in a manner consistent with the plan?

These sample watershed management scenarios capture the multidisciplinary, multiobjective, and multistakeholder aspects of watershed management. Wild watershed systems are enormously complex. The pursuit of knowledge of the interactions between climate, soils, water hydrology, plants, and animals at numerous scales in space and time will keep generations of scientists fully employed. Human cultural, economic, social, and technical systems obviously add substantially to the complexity. While scientists continue to study watershed systems, the management of the land and water does not pause. Best management decisions are based on the best available information and science, understanding that current and planned research will provide a basis for modifying these decisions in the light of future new knowledge.

As we have seen, a significant amount of the knowledge of the state of watershed systems as well as the dynamics of those systems is now captured in dynamic simulation models and as geographic information system (GIS) databases. Out of necessity, scientists have partitioned the study of complex interacting components of the systems into manageable (but still challenging) components. Because the movement of water is so well understood in terms of classical physics, we now have access to a stunning variety of powerful water flow simulation models. Vegetation and community succession models are only now becoming available, and are improving. Models predicting the habitat and populations of individual species are also available, but are typically developed on a species-by-species basis, because a general ecological theory of species niche requirements is not yet available. Climate models are now available, but watershed managers do not have the luxury of planning over many hundreds of years. The multiple weather models used to generate our daily forecasts are good only for a week into the future at best. Models representing local economies, urban growth patterns, traffic patterns and demands, crime patterns, and settlement patterns are also available and individually powerful. Unfortunately, the nature of disaggregated models seriously diminishes the applicability of the models for making land management decisions.

Currently, general-purpose dynamic watershed simulation models are available. They provide the ability to concurrently simulate several salient human and environmental processes. Modelers most often accomplished multidisciplinary simulation modeling by connecting two or more separate simulation models. For example, it might be useful to link a hydrologic simulation model to an ecology-based community succession model. This approach introduces software integration challenges that typically are addressed in an ad hoc manner reflecting the particular needs associated with the two models. Several groups have tackled the problem of model linking and integration in a more generic fashion. Two of the solutions are briefly reviewed in the next section, followed by a review of example location-specific models that integrate human and natural processes.

6.1 Sample Model Integration Environments

A number of existing approaches provide frameworks for integrating disparate simulation models. One approach is to recast existing (legacy) simulation models as simulation modules in a library of software subroutines or programming objects. The first system reviewed, the modular Modeling System, is an example of this approach. The second approach is to recast existing simulation models as separately running programs that, at run-time, interact with other running programs as a federation of simulation model programs. We will look at the Dynamic Interactive Architecture System as an example of this approach. Finally, an attrac-tive approach to many is to create an environment within which new modeling components can be developed with integration planned up front. The Spatial Modeling Environment is presented as an example of this approach.

Turning Simulation Models into Modules

The Modular Modeling System (MMS)[1] is a powerful contemporary watershed simulation environment that links a broad number of multidisciplinary models representing different aspects of land, water, and weather processes (Leavesley et al. 1995, Leavesley 1996). Its development was coordinated and led by Dr. George Leavesley of the U.S. Geological Survey (USGS). The MMS was created to facilitate the rapid development of watershed models. Watersheds may contain any number of different components, such as overland water flow, transpiration of vegetation, groundwater, soil saturation, weather/climate, streams, dams/reservoirs, insolation, and others. The MMS provides a library of software subroutines that model these different components, creating a set of watershed model building blocks. A graphical user interface (GUI) allows users to graphically arranging the logical connections among the watershed components to build a full watershed model. Once arranged, software associated with the building blocks is compiled (through the GUI) to create a single program that captures all of the processes. The resulting program is a simulation model, which runs the various interconnected processes using a common simulation clock. Run-time GUIs allow the visualization and capture of system state information.

The MMS began as a cooperative research effort between the USGS and the University of Colorado's Center for Advanced Support for Water and Environmental Systems (CADSWES). As MMS took shape, many national and international agencies and organizations expressed interest in the MMS

---

[1] MMS—http://www.terra.colostate.edu/projects/mms.html.

concepts. As a result, the MMS now contains model components developed collaboratively by a large number of organizations.

Integrating Separately Running Simulation Processes

The Dynamic Information Architecture System (DIAS),[2] is a powerful software environment developed by and for the Decision and Information Sciences Division at Argonne National Laboratory. This environment was developed precisely to facilitate a solution to the challenge of linking disparate multidisciplinary simulation models. DIAS allows software engineers to write simulation models that, at run-time, communicate with modified versions of legacy simulation models. The communication is two-way, which means that through DIAS, the different running models can exchange watershed state information with one another. DIAS was initially developed in support of a Dynamic Environmental Effects Model (DEEM)[3] funded by the Defense Modeling and Simulation Office (DMSO). DIAS has numerous support capabilities that accept data from various GISs and DBMSs (database management system) formats, has a built-in object-oriented GIS, and many internal simulation models. DIAS provides the software "glue" to hold together any number of discipline-specific environmental, economic, and social models useful in watershed modeling.

In Fig. 6.1 the one large and two small rectangles represent separately running processes. The large rectangle represents a core DIAS program. In this program, the "Event Manager" communicates with all of the model's objects (represented by the word "object" inside large ovals. These objects are DIAS compliant and are selected by a modeler from a "Frame Toolkit's" Object Library. The objects may contain all of the software code required for representing an aspect of the process being modeled. Alternately, they may accomplish their simulation modeling requirements by passing requests to "Process Objects," which make calls to computer programs (the small "External Program" rectangles) running simultaneously in parallel. Such programs are typically scientific models that were originally developed as stand-alone simulation models. By encapsulating the simulation software associated with these models with a layer of software that supports interprocess communication to the "Process Objects," the former stand-alone models can participate in a multidisciplinary simulation model. For efficiency, most additions to the growing DIAS library are written directly as internal DIAS simulation objects. The DIAS software represents a powerful and ingenious approach to the integration of multidisciplinary simulation modeling that will be necessary to fully use scientific models in watershed management decision processes.

---

[2] DIAS—http://www.dis.anl.gov/DIAS.
[3] DEEM—http://www.dis.anl.gov/DEEM/.

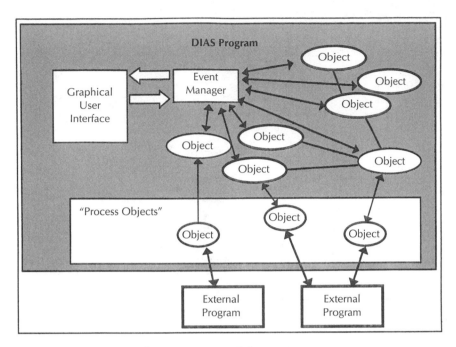

FIGURE 6.1. A conceptual representation of the DIAS environment

Frameworks for Creating New Simulation Model Modules

From a software engineering approach, getting disparate simulation models to interact with one another is best accomplished by building the simulation models within a predefined framework. An excellent example of this approach is the Spatial Modeling Environment (SME),[4] which is discussed in detail in Chapter 7. Figure 6.2 outlines the process for model development. None of the steps in this process involved writing computer programs in the traditional sense. The patch-based submodels were developed using a dynamic simulation modeling software environment called STELLA[5] in step 1. (The STELLA modeling environment is an example of a commercial product that makes it easy to specify simulation models through the development of algebraic and logic equations that describe how a system changes between time steps.) Step 2 involved the automatic conversion of the STELLA-generated equations into C++ computer language using the SME. The resulting software arranges for the STELLA-developed model

[4] SME—http://kabir.cbl.umces.edu/SME3/.
[5] High Performance Systems, Inc.—http://www.hps-inc.com/.

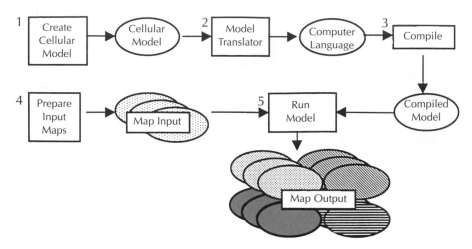

FIGURE 6.2. The Stella/SME model development process

to be run simultaneously within each landscape patch. To initialize the state of the system in these patches, GIS database development and analysis was conducted as needed (step 4). At run-time (step 5), these data layers were read into memory and the STELLA equations were applied repetitively (for each time step), resulting in landscape state changes over time that could be captured as map output and as tables.

Step 2 results in a simulation model module that can be added to a local library of simulation modules. A goal of the SME developer is to facilitate the development of modules within local libraries that can be easily shared with other SME users. Sharing can be very practical when developing new capabilities for which the software was designed. Recasting large legacy models in this environment is impractical, however.

6.2 Sample Multidisciplinary Models

An Everglades Environmental Model

A struggle between important stakeholder groups in South Florida has led to the development of a number of watershed simulation models. The unique Everglades environment has developed over hundreds of thousands of years, adapting to a continuous flow of water from the central areas of Florida through the Everglades and Big Cypress Swamp to the Gulf of Mexico. In response to water demands from growing South Florida human populations, much of the natural flow of water has been diverted to cities and agriculture. Environmental interests are struggling to slow and reverse the trend in wetland losses while growing human populations continue to demand more water rights. Both interests support the development of rather large simulation models in the hopes that acceptable land management plans can be better developed.

There are two basic components to the modeling approach: hydrologic and ecological. Water management scenarios are evaluated with overland, stream, canal, and pumping station models that predict the water flows and depths in the natural systems. Because water flow models are based on relatively straightforward and well-understood physics, these models can be quite precise and can be parameterized to represent the local system. Output from the water-oriented physical models then provides input for analysis of the long-term and cumulative effects on plant and animal populations. This analysis is accomplished via a simulation developed through a multiuniversity effort and is called the Across Trophic Level Simulation System (ATLSS).[6] Different components of the natural system are modeled at appropriately different levels of aggregation. Large animals that individually cover wide areas (deer, panthers, alligators, herons, and storks) are modeled at the level of the individual. Populations of smaller animals are modeled as populations, and the lowest trophic levels are captured in process models. Researchers separated by geography, organization, and discipline independently developed the different components of ATLSS in a manner that allowed the models to be integrated as a single system. ATLSS is a management-oriented system designed to assist in the analysis of alternative land management scenarios, and is based on the application of currently known science. The ATLSS staff was requested by the South Florida Water Management office to input the hydrologic constraints, as defined by the hydrologic model outputs, into ATLSS. The model generates a volume of reports on the impact of the proposed water management plan on the animals and plants of the Everglades and Big Cypress Swamp. This effort provides a benchmark for a high-end effort in the application of integrated scientific models to land and watershed management decision making.

Two Military Installation Environmental Models

The management of military installations requires a balance between the maintenance of a natural environmental setting and the operation of tracked and wheeled vehicles in a given watershed. The military requires soldiers to "train as they fight." This means that soldiers must not be required to muddy their thought processes in a war situation with any information or considerations that are not immediately relevant to survival and the meeting of tactical objectives. Soldiers are not encouraged, for example, to learn the native flora and fauna so that they might avoid harming a threatened or endangered species. When they cross a stream or river they

---

[6] ATLSS—http://atlss.org/.

must focus only on the current mission and the safety of themselves, their companions, and their equipment. Ideas of impacts on stream life, damage to root systems that impede erosion, and downstream water quality are of no consequence to a soldier in a war situation and are not part of their preparation for war. Management of the land upon which soldiers train is, however, extremely important because currently available training areas must be sustained to support training indefinitely. Management can only be accomplished in strategic ways that involve the scheduling of training in time and space. Two early multidisciplinary landscape simulation models were developed in the 1990s to help compare alternative strategic management plans for military installation management: one for Fort Riley, KS, and the other for Fort Hood, TX.

The Integrated Dynamic Landscape Analysis and Modeling System (IDLAMS)[7] was originally developed to help land managers and decision-makers at Fort Riley, KS, find cost-effective approaches to long-term land stewardship goals in support of the fort's mission. IDLAMS provided a computer technology that enabled land managers to integrate their planning process by identifying multiple land-use objectives and implementing trade-off analysis, evaluating the costs of viable land management alternatives, and incorporating "what if" scenarios into their decision making. IDLAMS integrated predictive models, decision support techniques, and a GIS that incorporated remote sensing and field inventory data. IDLAMS' user-friendly interface allows the resource manager to operate this predictive, decision-making, and planning tool without substantial GIS or computer modeling experience. IDLAMS could be launched from commercial-off-the-shelf (COTS) GIS software. Research scientists conducted IDLAMS research and development at Argonne National Laboratory (ANL) and the U.S. Army Construction Engineering Research Laboratory (CERL) with funding sponsored by the Strategic Research and Defense Program (SERDP).

At the core of IDLAMS are (1) a vehicle use intensity model, (2) a vegetation dynamics model that projects the status and succession of the vegetation communities, (3) a wildlife model, (4) an erosion model, (5) an economics model, and (6) a scenario evaluation module. An installation land manager begins evaluation by identifying the multiple objectives of interest to the fort. These could include objectives to minimize cost, minimize environmental impact, minimize erosion, minimize soldier travel time to training, and maximize the training effectiveness.

Alternative management scenarios that involved varying training intensities in time and space would then be evaluated by IDLAMS with respect to the objectives. A decision support module helps the land manager rank-order the alternatives by applying manager-identified objective trade-off information.

---

[7] IDLAMS—http://www.es.anl.gov/htmls/idlams.html.

The IDLAMS model was adapted to a very different ecosystem from the Fort Riley tallgrass prairie when it was modified for the forested Fort McCoy ecosystem of central Wisconsin. Because the ecological processes are so different from site to site, environmental modeling for watershed management must reflect a different set of concerns and processes at different locations. IDLAMS responds to this fact by establishing a general framework for landscape simulation modeling, but allowing for the rapid removal, replacement, and update of its decision support system components. This means that scientists and software engineers must be available to adapt IDLAMS to other locations.

A second spatially and temporally explicit model developed for military installation land and watershed management is the Fort Hood Avian Simulation Model (FHASM)[8] (Trame et al. 1997). FHASM was developed within the SME framework. It focused on the impact of alternative land management practices on the populations of two federally designated endangered bird species. Like IDLAMS, the landscape was represented as raster GIS data layers—creating an array of square landscape patches. Dynamics of the landscape were divided into four major submodels: avian ecology, plant community, management decisions, and impacts.

The value of FHASM, and other models developed with SME, is that while each landscape is unique in its management challenges, important processes, and habitat definitions, software such as SME allows resource managers to focus their energies on capturing knowledge salient to that system. SME, in turn, provides a computational environment that removes the burden of programming the low-level software for user interface, data input and output, dynamic simulation, and distributed execution.

This chapter provided an overview of current approaches to multidisciplinary simulation modeling by looking at two model development environments (MMS and DIAS) and three multidisciplinary simulation models (ATLSS, IDLAMS, and FHASM). While these examples demonstrate that multidisciplinary models can be and are being developed to address management concerns, they are challenging to develop and can be difficult to operate. So far, these types of models are proving useful only in well-funded situations. Multidisciplinary models are now assembled in an ad hoc fashion to meet the particular needs of a client.

---

[8] IDLAMS—http://blizzard.gis.uiuc.edu/htmldocs/HoodModel.

7
Creating New Models

For some watershed simulation modeling purposes, simulation models must be created and not merely adapted and/or parameterized. This is particularly true for ecological simulation modeling. Instead of adopting a simulation model that is parameterized for a local watershed, simulation model development environments are used to develop a locally explicit model. Figure 7.1 compares modeling environments and models. Software models, such as those discussed previously in Chapter 5, offer the user a complete mathematical formalization of a set of watershed processes. Associated with the model are data import and export, data storage and retrieval, and model execution interfaces. Various types of data output and data visualization capabilities are often provided for internal and/or external output analysis. Through visualization and analysis of output, decisionmakers become better informed with respect to potential consequences of alternative watershed management decisions. Modeling environments exist to allow the efficient development and use of models. They typically provide user interface, data storage and retrieval, libraries of submodels, and simulation modeling primitives to which the user adds the actual modeling equations and the initializing system state information. In this chapter we explicitly look at software options targeted for each of these purposes.

7.1 Geographic Information Systems–Based Approaches

Computer-based geographic information systems (GISs) were first introduced in the 1960s; by the later part of the next decade, good commercial GIS software was available. The Harvard Laboratory for Computer Graphics and Spatial Analysis, under the direction of Howard Fisher, introduced a series of GIS software products including SYMAP (mid-1960s) and ODYSSEY (mid-1970s). Since then, numerous commercial enterprises and government agencies have offered a variety of GIS products. Some of those efforts continue to play a role in the GIS industry today. The Environmen-

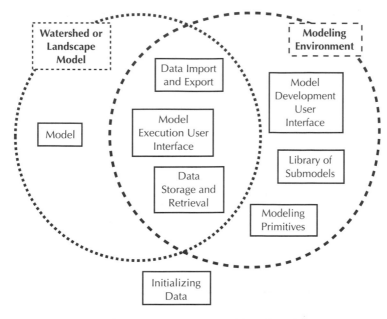

FIGURE 7.1. Distinctions between computer model and modeling environment

tal Systems Research Institute (ESRI)[1] has led the commercial GIS industry for nearly three decades and began with experiences grounded in the early Harvard efforts. The most notable of the U.S. government–funded efforts is the Geographic Resources Analysis Support System (GRASS),[2] now housed at Baylor University. The GRASS effort spawned a number of commercial efforts as well, including the commercial GRASSLAND product.[3]

Modern GIS software offers the watershed manager the ability to formalize the state of the landscape system. It allows us to capture current and historic information about the location of things. Data sources include historical maps and photographs, satellite and high-altitude imagery, Global Positioning System (GPS) surveys, and scientific analyses of all aspects of the land and water. Powerful visualization and data manipulation tools allow for the derivation of secondary maps (e.g., slope, elevation, and watershed boundaries from digital elevation maps). Powerful spatial statistics and mathematical analyses allow us to thoroughly explore our formalizations of what is and has been on the watershed.

The modern GIS does not afford us good opportunities for capturing our theories about how that watershed works. The role of GIS in the develop-

---

[1] ESRI—http://www.esri.com.
[2] GRASS—http://www.baylor.edu/~grass.
[3] GRASSLAND—http://www.globalgeo.com.

ment of dynamic simulations of human and natural processes has been primarily relegated to the input and output of spatial data to initialize external simulation models and accept output from these models for visualization and analysis purposes.

However, modern GIS environments can be used, on their own, to develop dynamic watershed simulations. For example, most modern GISs provide some form of map calculator allowing a new map to be created that is based on the mathematical analysis of one or more input maps. It is possible, for example, to create a hydrologic overland water flow simulation with a map calculator. The depth of water in each grid cell is a function of the depth and flow of water in the previous time step and the amount of new rainfall since that time step. Using a map calculator, finite difference equations can be developed to update depth and flow information. The equations are readily applied to every cell repeatedly, resulting in a time series of flow and depth. Similarly, the movement of populations of animals, plant seeds, or environmental pollutants can be captured in terms of the GIS map calculator. There are two drawbacks. First and foremost, the map calculators are designed to run only one step at a time. A simulation that involves many time steps must therefore involve the loading of the map calculator program, the current map data, the computation of the next time-step state, and the output of that state. This is very inefficient and can require many times more computational horsepower than necessary. Second, the GIS does not provide an efficient environment for developing, storing, mixing, and executing these models. Advantages to doing simulations within the context of a GIS are the powerful spatial analysis capabilities that can be employed during simulation operations.

7.2 Introductory Dynamic Modeling Software

For many watershed managers and students, the possibility of personally using computers to develop simulation models for comparing alternative land management scenarios can be intimidating. Most of us have used computers for word processing, e-mail, and perhaps spreadsheets and statistical analysis. But the formalization of concepts and understandings of watershed dynamics must not be reserved for trained computer scientists. While it is true that computer scientists have helped develop many of the software models reviewed in this book and software engineers will be intimately involved with future simulation modeling, it is important that the managers working with programmers (or their products) have a working understanding of the processes to be modeled and the approaches to capturing system dynamics in software. To assist, there are a number of different types of simulation environments for exploring how to gain an understanding of system processes and capture them formally in a model as system dynamics. We will consider software simulation options that are

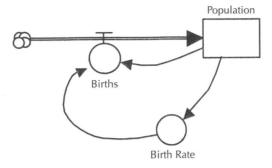

FIGURE 7.2. Stella example

not spatially explicit and those that consider the geographical placement of watershed components.

Stella, created by High Performance Systems[4] and PowerSim[5] provide powerful, flexible, and easy-to-learn simulation model development environments that are not spatially explicit. Each provides a graphical user interface (GUI) that allows for the rapid capture of the relationships among system components. Figure 7.2 is a sample STELLA model showing the key components from which dynamic simulation models are constructed. The rectangular icon represents a system state value, in this case, population. It is the collection of state variables in a model that defines the system at any given time. For Stella models, a fixed time step is chosen for performing simulation calculations. The circle associated with a double line controls the increase in population. A useful metaphor is to view the rectangle as a reservoir, the horizontal double line as a pipe through which water flows, and the circle on the pipe as a valve that controls the rate of flow. Circles that are not attached to pipe lines simply hold equations or variables. The arced lines indicate that the circles at the arrowheads are functions of the icons at the other end of the arcs. Hence, in this example, "Births" is a function of "Birth Rate" and "Population." "Birth Rate" is simply a function of "Population." The modeler must then enter equations that use the indicated inputs. In the case of Stella, double clicking on the "Births" icon reveals a GUI that allows the modeler to provide a suitable equation using both inputs ("Births" and "Birth Rate"). Stella simulations proceed by computing all equations and then updating the state variables. The Stella approach to simulation modeling makes use of difference equations that are written using fundamental-level algebra and basic logic. The equations need not be continuous. If a suitable equation is not available, is too difficult to describe

[4] High Performance Systems, Inc. http://www.hps-inc.com/.
[5] PowerSim. http://www.powersim.com/.

in mathematical terms, or is directly supported by measurements, the modeler can describe the relationship between two variables by drawing a graphical relationship within the Stella interface. For example, if birth rate is a function of population, the Stella modeler can specify that a graphical relationship will be used instead of an equation. Stella presents a graph with the dependent and independent variables and the modeler traces a graphical relationship between the two. These modeling environments allow users to rapidly formalize their understanding of a system's dynamics. This type of model development environment is very accessible, requiring knowledge of the system to be modeled and basic understanding of high-school algebra. Hannon and Ruth (1994, 1997) have published enjoyable texts that provide simple demonstrations of the Stella model development environment for ecological, environmental, and economic systems. Once an individual develops an understanding of how a system works, the computer then reflects back the consequences of that understanding by repeatedly applying the modeler-specified dynamics to generate a time-series representation of the state of the system.

Stella, PowerSim, and related environments are not practical for exploring the role of the spatial structure of a watershed system. Geographic information systems explicitly model the state of such systems, but are weak in their ability to simulate a system's behavior over time. For exploring the implication of space in dynamic systems, a number of new environments have recently become available. StarLogo,[6] a product of the Massachusetts Institute of Technology's Media Laboratory, and EcoBeaker[7] are powerful educational tools for exploring spatial dynamics. These simulation environments have friendly and flexible GUIs that allow modelers to focus on the description of the interactions among agents, and between those agents and the patches of land over which they move. These patches can similarly interact with neighboring patches. Both systems provide simple programming languages that modelers use to describe the interactions of interest and simple buttons and sliders for creating model-specific user interfaces. Sample models are provided with each system that allow the user to begin exploring predator–prey interactions, traffic, disease and epidemics, fire simulation, vegetation growth, foraging, social interactions, and various types of competition.

Figure 7.3 shows a portion of a sample StarLogo-based pollen-spread simulation model. Here, a single corn plant is releasing pollen under user-specified wind speed, wind direction, temperature, and humidity conditions. The graphical view is from overhead with north to the top of the image. Pollen (bright spots) moves to the east (the WindDir slider is set to 90 degrees) and is randomly moved perpendicular to the wind direction and

---

[6] StarLogo—http://StarLogo.www.media.mit.edu/people/StarLogo/.
[7] EcoBeaker—http://www.ecobeaker.com/.

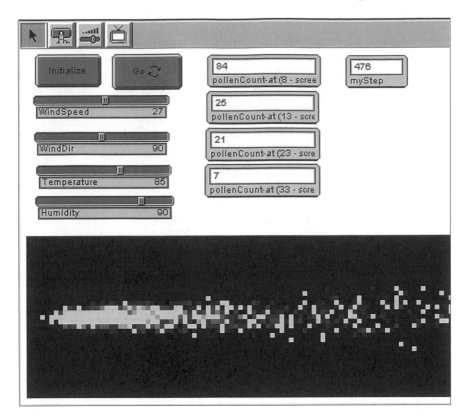

FIGURE 7.3. StarLogo-based pollen-spread model

also in the up–down axis. If the pollen drops below a certain height while still viable, that location records the event. After running the simulation, a pollen deposition pattern emerges. Fewer than 100 lines of StarLogo code capture the dynamics. StarLogo automatically compiles code and provides a simple interface for running models. It is perfect for exploring simple ideas in spatially explicit simulation modeling.

Software environments such as StarLogo and EcoBeaker are powerful, flexible, and inexpensive. They are perfectly suited for exploring ideas about watershed processes.

7.3 Power Dynamic Modeling Software

For development of large, complex, and computationally intensive spatially explicit simulation models, programs such as StarLogo and EcoBeaker are not adequate. These programs are not well suited for working directly with digital maps in a GIS or interacting with any other spatially explicit simulation models. They are made to handle relatively small simulation programs—making no provision for easy management of more than a few hundred lines of software. They also do not provide for the distributed processing of very large simulation models that could benefit from multiple processors or multiple computers. For more serious simulation modeling, one must turn to more open-ended environments oriented to software engineers. Unfortunately, more powerful, yet reasonably priced, software is not yet available in the marketplace. However, a growing number of alternatives developed by various government laboratories are available. The Spatial Modeling Environment (SME)[8] and SWARM[9] provide examples.

The SME marries simulation modeling software such as STELLA to a powerful simulation execution environment. The SME facilitates the simultaneous execution of STELLA-like models for each grid cell associated with a raster GIS database. State variables in the models are initialized using information in GIS data layers. The SME adopted the model–view–driver architecture approach displayed in Figure 7.4. Modelers are not expected to be software programmers, encouraged only to develop the STELLA-like models that will be run in parallel—accommodating each patch in a watershed grid. The SME models are written using a simple convention that allows a cell state variable to be a function of not only the variables associated with the current cell but also the variables of neighboring cells. Unit models are translated by SME Module Constructor software that translates the various modeling environment equation files into a common Modular Modeling Language (MML). A library of such translated models can be built up and maintained for future use. To create a spatially explicit simulation model, the modeler identifies model components (submodels) from their library, matches variables where appropriate, and translates the MML code into C++ using the SME code generators.

Other SME options allow for the integration of channel flow process models and point models. The SME-generated models can read and write various GIS data formats and tables of data in different formats, and can generate variable graphics and maps during simulation runs. Because SME is written in C++, it is possible for a software programmer to link SME code to other simulation models.

While SME works very hard to hide the development of software programming language code, SWARM empowers the software engineer. SWARM was developed at the Santa Fe Institute, Santa Fe, NM, to support

---

[8] SME—http://kabir.cbl.umces.edu/SME3/.
[9] SWARM—http://www.santafe.edu/projects/swarm.

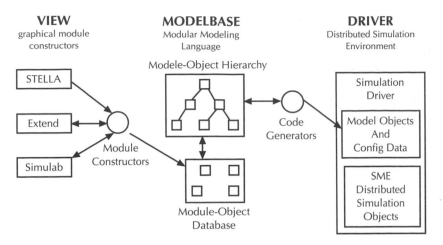

FIGURE 7.4. SME's Model-View-Driver approach

the exploration of artificial life to provide insights into the development of societal behaviors. Like STELLA, EcoBeaker, StarLogo, SME, and others, SWARM provides the modeler/developer with modeling support software that takes care of simulation time, GUIs, and various output visualizations. And, like the others, it uses a programming language that allows the modeler to capture their understanding of the dynamics of the spatial system. But, unlike the others, SWARM does not provided the developer with a comfortable GUI environment for developing the models. SWARM provides its capabilities in the form of object libraries that a software engineer accesses as extensions to a powerful programming language called Objective C—an object-oriented set of extensions to the popular C language.

SWARM encourages the programmer to develop "experiments" within its modeling approach. SWARM provides for the passage of simulation time, visualization of the system state during simulations (two-dimensional images, tables, charts, and pen-plots), and the probing of model components. SWARM is a tool for software programmers and can be useful in the development of decision support systems.

SWARM is not an acronym, but a modeling approach that involves hierarchical groups of related objects—swarms of objects. A swarm, in this context, is a group of objects that interact with one another and are kept in synchronization with one another by sharing the same simulation clock. One or more of those objects may themselves be swarms and this relationship can continue hierarchically through swarms of swarms of swarms.

The simulation modeling environments described above each have different strengths, requirements, and uses. As one moves from an exploratory mode toward the development of serious management support models, it becomes increasingly important to involve more players in the modeling game—each with a different set of specialties. The next chapter explores approaches to success in multidisciplinary simulation modeling.

8
Coordinating Large-Scale, Interdisciplinary Watershed Modeling

Recent advancements in computer hardware and software have made it possible for small land management offices to develop and run complex ecological models. This is not yet common because the opportunity is relatively unknown, the development of serious models for supporting land and water management requires a multidisciplinary and multistakeholder effort, and often the models must be developed to address local requirements. Because natural systems are far more complex than what can be captured in a single general ecological simulation, models are generally developed for particular watersheds and for particular management purposes. A growing number of local, state, and federal land management offices are seeking to develop such models to assist with land management decisions. Successful development of such models requires the effective creation, management, and collaboration of interdisciplinary teams. This section provides a road map by which dynamic, spatial, ecological models can be successfully developed by large interdisciplinary teams. Model development team members as well as team leaders will benefit from using the model development requirements and stages described here.

A dynamic, spatial, ecological simulation model is a computer-based simulation that evolves a watershed through time. Mathematical, logical, and stochastic processes, defined and created by the modeling team, drive the simulation. These processes are functions of the state of the system; their output is the continual redefinition of the system. The system state is seeded by the modelers with a "picture" or "snapshot" of the system at some real or artificial start time. Seeding may use various forms of input data including raster images, vector data, point information, and object status and location.

Stepping a simulation through fixed time steps may accommodate the passage of time. Alternately, simulation processes can be dynamically scheduled within the simulation to allow different processes to occur at more appropriate intervals or only at times that make sense for the simulation. For example, the drainage of storm water need not be computed during dry periods, but should be "turned on" at the outset of a storm event. Many models use a single time step that is carefully chosen to (1) make sure that the dynamics of the most rapidly changing events are captured and (2) make sure the model can be simulated with overall speed that is acceptable.

8.1 Modeling Steps

The design and development of a modeling environment must pay attention to certain principles, objectives, and approaches to modeling in addition to considering the theoretical foundations of the associated discipline. The opportunities provided to a modeler in a software environment are important in the process of model construction. Each model creator develops a set of steps that prove to be efficient in the development of useful models. Software that reflects the modeling steps or processes is preferred. Overton (1977), for example, lists the following required steps:

1. List the model objectives.
2. Identify submodels and subobjectives.
3. Construct and validate submodels.
4. Assemble the submodels into the complete model and validate.
5. Attempt to address the questions identified in step 1.
6. Examine the general behavior of the model; identify behaviors of interest.
7. Conduct sensitivity analyses; identify the structure and parameters that are causal for the behaviors of interest; validate those causal structures and parameters.

It is important to identify the objectives for any project. Objectives are not always commonly shared and large groups can easily spend significant time just agreeing on a set of objectives. In fact, this step (step 1) can easily be expanded into a half-dozen steps that move a group through human dynamics that finally involve a group educational process on the details of the watershed management challenge. If a group decides that development of a simulation model is required, that group addresses Overton's second step. If the system is sufficiently complex, it will be important to break the intended model into components called submodels (step 2). An automobile engine is composed of subsystems (carburetor, exhaust system, crankcase, fuel system, electrical system, and the cylinder assembly), which in turn are composed of smaller subsystems. It is possible that the submodels in the simulation model under development should also be further divided. A hierarchical organization of a large effort makes it possible for individuals to focus time, energy, and intellect on the different components at the different levels in a team approach. A watershed group might discover in this process that they have collectively moved their individual and group understandings of the system to a level where going through with model construction will be unnecessary. Step 3 begins the actual model construction phase, and it can occur simultaneously on each submodel. When the components are ready, they can be assembled (step 4). Depending on the number of subcomponents, assembly can be challenging, especially if the different development groups used different modeling approaches, data definitions, and/or assumptions. Using the completed model, the original question for which the model was developed can be addressed (step 5). Models can offer more insights than those sought and it is useful to experiment with the model to test responses to different combinations of inputs (step 6). Overton's final step (step 7) is to analyze the model with respect to its sensitivity to the uncertainty in model inputs and assumptions. This step helps to establish an appropriate level of confidence in the model and also to identify the importance for collecting further information about the watershed.

A slightly revised set of steps is listed and developed below in detail. These have been developed and used in classroom-based, multidisciplinary, spatially explicit simulation modeling.

1. Identify objectives and constraints.
2. Develop overall modeling constraint decisions.
3. Conceptualize full model.
4. Develop submodels.
5. Develop full model.
6. Iteratively test and debug.

Step 1—Identify Objectives and Constraints

Identify the End User

Identify the end user of the model and then develop a precise statement of the user's objectives and requirements. Any project initially requires an identification of the objectives, followed and modified by constraint identification. In the design of a building, interplay between the artistic goals and engineering possibilities is necessary. For simulation model design and development, the following must be addressed:

- **Characterize the end user.** This step is easy to overlook; consequently, modelers often develop models as if they were the end user. Characterize the intended user or user community with respect to their academic and professional experiences. What skills will they bring to the effort? How often will they use the model? What learning curve will be tolerated? The end product must match the capabilities of the end user and the frequency of their anticipated use. Long learning curves are more acceptable if the product will be used regularly.
- **What decisions will be made?** What does the end user require of the final model? What does the identified end user expect from the modeling

effort? Answering these questions requires a significant amount of communication and conversation between the modeling staff and the end user. The answer is not always immediately apparent to the end user, but instead may evolve over time and relies heavily on open and honest communication between the modeler and the user. The modeler(s) must regularly back up from the intense and focused modeling effort to reevaluate the role of the anticipated product in the bigger picture.

- **How accurate do the output requirements need to be?** Perhaps it is sufficient for the model to be able to generate relative or suggestive output to show trend directions. Alternatively, end users might require highly accurate data useful for fine-tuning land management scenarios.
- **How much funding is available for the effort?** This is perhaps the single most important consideration, as all the answers to the other questions will be heavily influenced by the question of funding.

Identify the Available Resources

In particular, the following questions must be asked:

- **What data are available?** Any ecological model is only brought to life when married to an initial system starting state. Can this state be constructed with readily available data, or must the state be defined through expensive and time-consuming fieldwork?
- **What expertise is available?** If you have the "right" people, model development can proceed with the greatest efficiency. If expertise must be acquired, will team members need to be educated in the missing expertise or can the expertise be temporarily borrowed or purchased?
- **What applicable models already exist for the system under consideration?** Is this a first effort in building models for the system under consideration, or do models already exist that have captured useful data and or system dynamics?

Data include spatial data defining the state of the system under consideration and may take the form of raster maps, satellite imagery, vector maps, polygon maps, and/or point data (representing data sample points). System interaction rules will be added to the state of the system. These rules define another data need. What literature, experience, or completed modeling components already in software are available to the modeling team?

Identify Tool Availability

The tools of modern ecological modeling activities are in the form of computer hardware and software.

- **What computer hardware is available?** Virtually all location-specific ecological models run on digital computers. What computers exist locally? Are they available for the project? What hardware might be accessible via a local or extended network? What bandwidth is available for inter-machine communication?

- **What software capabilities are available?** Software is tied to licenses, particular platforms, and versions of hardware operating systems. Software must be identified with respect to availability, associated hardware, and intersoftware-data-sharing capabilities.
- **What costs and benefits are incurred in hardware and software use?** Perhaps the best available hardware and software will simply allow a model to run a few seconds (or minutes) faster than alternative hardware and software that cost half as much.

Consider the Level of Effort Possible

Addressing this issue requires consideration of the following questions:

- **How much time can each participant provide to the effort?** Regardless of how fast, efficient, and proficient an individual, team, or organization is at accomplishing a task, availability to accomplish the task is critical and easy to overlook.
- **What time frames are available for interteam coordination?** For interdisciplinary efforts, viability of group members must be judged with respect to how much time the members can be available to each other.

Answers to the above questions create a set of limitations within which the modeling activity must take place. Once answered, the team will consider whether there is sufficient space within the range of possibilities to pursue the development effort further. The modeling manager(s) will develop understandings of trade-offs among fund expenditures on the various personnel, hardware, software, and other modeling inputs. Discussions with the sponsors and potential end users at this point will help firm up the limitations with respect to final funding decisions. This will then be followed by firm personnel assignments and project expectations jointly developed by the project collaborators.

Step 2—Develop Overall Modeling Constraint Decisions

Now the process of designing the final model begins. Based on personnel, funding, and hardware and software decisions, the modeling team must first identify the fundamental framework of the target model. This includes up-front decisions with respect to potential model components, modeling time steps, spatial resolution, spatial framework (raster, vector, point, or objects), time frame, principal outputs, and fundamental purpose of the exercise. Each factor is reviewed individually, though in many cases, decisions are not independent of each other.

Identify Potential Model Components

A broad array of model components is conceptually available to the watershed modeler. Decisions made in the choice of hardware and software, however, will significantly limit the options available. Generically, potential model components to be considered include the following:

- **Watershed patches.** Watershed characteristics are now recognized to be important variables in models that capture the movement of individual organisms, structural changes in ecosystem boundaries, and the movement of air and water. Dividing the watershed into grid cells, hexagons, or irregular polygons can best capture some spatial information. For the sake of simplicity, most models choose to operate using data in only one of these formats and at a fixed resolution. (The resolution is often chosen with respect to the operational time step of the modeling system.)
- **Linear objects.** Some spatial structures are most efficiently captured as linear items. This includes streams, rivers, and most man-made structures, such as roads, fences, buildings, and parking lots. Accommodating both linear objects and watershed data stored in raster or hexagon format is complex, but often unavoidable.
- **Discrete, mobile objects.** If individuals, groups of individuals, vehicles, or individuals of an endangered species are key components of the model, they must be represented as discrete, mobile entities. Such objects must be able to "disconnect" from locations in the watershed space and "reconnect" in some adjacent space where they will interact with that environment.

Identify Potential Model Interactions

The possible model components have myriad potential interactions to consider. There are enough possibilities to fill another book and they cannot possibly be given sufficient consideration here. The following list suggests only the breadth of potential model interactions. The modeling team must identify the *best model interactions for capturing the system under consideration*.

Raster GIS interactions.

Classes of traditional Geographic Information System (GIS) interactions have been developed and discussed by Tomlin (1991). Sample GIS operations include

- **Simple location-by-location overlays** that can be expressed with mathematical equations using maps as variables. Examples: (1) finding locations that meet certain local criteria, (2) finding correlation or potential local impacts related to a proposed change on the watershed, and (3) transforming a set of maps (e.g., slope, aspect, soil characteristics, rainfall) into new interpretations using mathematical relationships (e.g., soil erosion potential based on the Universal Soil Loss Equation).

- **Near neighborhood operations** that compute output as a function of the state of small regions surrounding each map location. Examples: (1) Computing slope or aspect as a function of the elevations surrounding each map location, (2) determining direction of flow of some quantity (e.g., water or air), and (3) identifying areas where information changes rapidly (edge detection).
- **Cellular automata interactions.** A two-dimensional cellular automaton is a surface that changes over time as a result of equations generating the future state of the locations (cells) in the system as a function of the current state of that cell and its four (or eight) neighbors. This approach was first popularized by Conway's "Game of Life" (MacLennan 1990). For exploring the qualities and characteristics of the cellular automaton approach, cells are often assigned to a limited number (<256) of states. Relaxing this to accommodate a large number of floating point stocks results in the type of cellular models used by Costanza et al. (Costanza et al. 1992, 1993). Here dynamic ecosystem models with fixed time steps are run simultaneously for a number of land parcels arranged in regular grid cell arrays. Each cell is treated as a homogeneous system that can be influenced in each time step by its state and the states of its neighboring cells.
- **Vector GIS interactions.** Watershed information stored as polygons and linear features can also represent objects that interact and move. Examples might be traffic flow patterns along roadways or the hydrologic behavior of stream or river networks during unusual storm events. Entities such as cities, parking areas, private land, or stable ecosystem regions are most efficiently stored as polygon data elements and can be conceptually easier to model as distinct entities.
- **Interactions among mobile objects.** Some modeling efforts require the recognition of distinct entities that can move across the watershed. Examples might be individual members of an endangered species, vehicles moving about a watershed, or a group of such objects that always remain in close geographical contact.

Interactions can occur as mixes or hybrids of these broad classes as well. An animal modeled as a mobile object must interact with water found in streams that are modeled as vector entities and vegetation that might be modeled as a component of a cellular model fixed in space. Entities can communicate with each other through sounds, pollens, propagules, pheromones, and waste gases that move through the modeled space. Clearly, the range of potential interactions is quite large. Generally, the available software will severely limit the options of the local modeling effort.

Time Frame

Once the purpose of the modeling effort has been determined, it should be relatively straightforward to identify the time frame within which the simulations will be conducted. The modeling team must base this decision on the anticipated length of time that a management decision might affect the future. This must be tempered with educated guesses about how rapidly the predictability of the model will decay over time. Like the weather reports on the evening news, which are based on expensive weather pattern models, the longer a simulation runs, the less confidence one has in its predictive capability. There may be exceptions to the time–confidence issue. For example, spatial models may show stability at a gross scale, but show apparently random output at a detailed scale; that is, an overall pattern persists among simulation runs, but the details of exactly where the patterns are located may change with different runs of the simulation. In general, the answer for the target time frame comes directly from the requirements of the end user.

Time-Step Considerations

Within the identified time frame, the model simulation will proceed from some starting time to the target simulation end. How will the simulation proceed through time? There are three basic aproaches: fixed, variable, and event driven.

- **Fixed.** This is conceptually the most simple, but functionally the most limiting. The simulation proceeds through time at a fixed rate (dT). The state of the system feeds into the model mathematics and logic to generate the system state for the next time step, which is always a predetermined fixed time into the future. The advantages are found primarily in the simplicity to the modelers. All equations are generated with respect to a known dT. Computation of time passing is neither necessary nor appropriate. Disadvantages are found particularly when the model must consider activities that operate at radically different rates. The growth of a plant may be modeled appropriately with a time step of a day, or even a week or longer. The flood that washes the plant away may require a simulation time step in the range of seconds. Choosing a dT that reflects the plant's behavior may not catch the action of the plant being washed away. Choosing a dT that will capture the storm may result in a simulation that never completes on available hardware.
- **Variable.** Another solution is to use a variable time step. This method may take two basic forms. First, there might be a single variable time step for the entire model. The time step is initialized to a long dT, which will be dynamically modified depending on the rates of changes detected within the model. When the storm hits, for example, the model will detect that fast-changing activities are taking place and will have the ability to rerun at a shorter time step. The second approach is to set different fixed time steps to different parts of the model. This approach provides some computational relief while maintaining the relative simplicity of fixed time steps.

- **Event driven.** In this approach, there are no explicit preset time steps; they have been replaced with a system calendar that schedules events. The plant submodel may execute and then schedule itself to be updated at some later time based on its rate of activity. The storm submodel may be scheduled to run only when a storm is actually scheduled to take place. While it runs, it may schedule the plant submodel to run much more frequently during the storm. This approach is most attractive for models that must make the best use of available computing resources, but it is the most time consuming from the modeler's perspective, as it requires significantly larger simulation models to be created.

Again, the available hardware and software will likely limit the number of options. When there are options, the modelers should weigh the costs and benefits of using the different approaches.

Spatial Resolution

Finally, the resolution of the spatial representation of the environment to be modeled must be considered. As the resolution of time discussed above is important, so too is the resolution of space as used in the models. Entities on the watershed that move or affect their neighborhood, should move or transmit their effects in a single time step no further than the dimensions of the spatial patches. For example, if a predator or a moving vehicle moves 100 units of distance in a dT over a terrain that is resolved to 10 units, that entity will not affect or be affected by intervening terrain. Instead, the entity will appear to "warp" through space avoiding any obstacles or opportunities in its path. Hence, fixing a dT for the entire model, or for portions of the model that accommodate movement, directly affects the resolution of the salient terrain features. Terrain resolution schemes can be categorized as follows:

- **Fixed.** The terrain features assume a fixed resolution, which can be constant across the watershed representation. A regular array of square grid cells or hexagons is commonly used in spatial simulation environments. A regular array has the advantage of being conceptually simple, as models need not modify their behavior based on different or changing resolutions.
- **Hierarchical.** Models that simulate activities that occur at different spatial resolutions may adopt a spatial data structure that maintains information in a hierarchical manner. Each cell or hexagon can be iteratively decomposed into increasingly smaller components. Movement of large entities (weather systems, flocks of birds, clouds of spores or pollens) can move rapidly across the system using relatively long time steps and large spatial patches. Entities that are smaller (individuals or vehicles) can operate at smaller time steps and smaller patches. In this scenario, data are maintained simultaneously at varying scales.

- **Variable.** In the case of large objects that move slowly across the watershed (ecosystems, roaming herds of ungulates, or populations of invading species), there may be the desire to maintain the entity as a single whole while retaining the detailed spatial structure defining the dynamically varying extent of the entity. This type of operation requires maintenance of the spatial extent at a fine resolution, but simulation of the object dynamics at a grosser resolution.

As is true with other considerations discussed above, the modeling staff will generally find their options predefined by the software and hardware limitations.

Step 3—Conceptualize Full Model

Actual model conceptualization can now begin, with participation from the full modeling team. At this point, the team has identified all of the constraints within which they will conduct their modeling exercise. They have identified overall goals and objectives, user expectations, and available expertise within the group. This final step before the individuals can begin to work on the details of the model should result in

1. Identified subcomponents of the desired full model
2. Inputs required by each subsection
3. Model initialization requirements
4. Available, but unused, model outputs

Full model concept development begins with a focus on the model output requirements. These requirements will most often take the form of a time-series output showing the status of something within the model. This might be the health of ecosystems, the state of an endangered species, property values, soil depth, land-use patterns, or the positions and paths of vehicles or individual organisms. In all cases, some particular set of output requirements must be recognized and must drive all subsequent decisions. For each of these outputs, the group then identifies what factors directly influence the state of that output. For each of these factors, they conduct a conceptual sensitivity analysis to help identify factors thought to be most important for the determination of state changes in the outputs under consideration. Information on the important factors is recorded to assist in immediate and subsequent model refinement. Modeling exercises tend to initially focus on variables with a wide variety of sensitivity characteristics. The modelers must seek to identify those model components, variables, and data requirements that are most critical for computing correct model outputs; that is, model outputs are sensitive to the values of inputs, but more sensitive to some. During model development, decisions to include and develop model components and data inputs must be based on a professional sense of the relative importance among the competing opportunities.

Iterate through this process until the group is comfortable that the conceptual model is sufficiently complex to answer the primary questions, but can still be accomplished within the constraints already identified.

In summary, the steps required to complete the full model conceptualization are

1. Identify and discuss the primary output desired. Begin with the questions that will be asked of the completed system.
2. Discuss the model inputs that might be required for generating accurate state changes in the primary state variables.
3. For each of the potential inputs, conduct a conceptual sensitivity analysis to help prioritize the potential inputs.
4. Iteratively repeat the previous two steps for each of the important inputs.

The modeling group has now assembled a rough conceptual model with a notion of the relative importance of identified components. This conceptual model was generated as a team effort based on team-identified model development constraints.

Submodel Identification

In this stage, the development team assigns to themselves various components of the conceptual model. Submodel development teams or individuals will then take their individual assignment, which is understood within the context of the larger needs, and develop a portion of the final model. They will retain "ownership" of their full model subcomponents through the life of the project. Hence, this splitting of the full model must be accomplished with care and must consider

1. Team member expertise
2. Team member availability
3. Team member learning requirements

This stage often proceeds smoothly because the capabilities of the team participants usually become apparent in the model conceptualization phase.

Submodel Requirements Identification

Submodel teams (possibly consisting of single individuals) now begin their modeling efforts. This effort starts with further conceptualization of a portion of the model with the immediate objective being external model requirement statements from the submodel teams. This process becomes an iterative conversation among submodel groups. Through identification of submodel input requirements and submodel output possibilities, the teams work toward interteam design and development agreement. The steps are

1. Subteams identify external input requirements.
2. Subteams identify potential outputs.
3. Subteams work toward agreement about submodel inputs and outputs.

The agreement must identify publicly available state variables, variable units, and variable resolution. The debate will likely proceed through several iterations before consensus can be found. It is important that this agreement not be taken lightly. If a subgroup works hard to develop a submodel based on the understanding that a key input will be available, the later unavailability of that input can severely weaken the cohesiveness of the full working group.

Model Initialization Requirements

As a special case of the identification of submodel requirements, model state variables must be initialized externally with a working picture of the system at time step 0. Depending on the approach to stock variable simulation, external data sources, such as raster and vector maps, site description tables, entity state descriptions, and external model output (e.g., a global climate model), will be required to seed the model's state variables. This effort can be as time consuming and difficult as the development of the model rules and equations. Team members will be assigned to this effort and will similarly debate and establish working agreements among the submodel teams.

Step 4—Develop Submodels

The identification of submodels, submodel development teams, submodel interaction requirements and expectations, submodel initialization requirements, and time frames for completion of the submodel components have all been completed. Submodel development teams can now focus independently on the further design, refinement, debugging, and sensitivity analyses of the submodel components themselves. Note that if the submodels are too unwieldy, they must be divided into manageable pieces using the procedures described above for the full model. Indeed, very ambitious models will require several levels of partitioning before individuals are allowed to focus on development of the cause–effect mathematics of the model.

The submodel development process is not described in any detail here because there are no defined steps that work well for all individuals or models. This is the area where individuals can be most creative in the model and the modeling process.

There are, however, a number of activities and objectives that must be considered during the submodel design and development. We will group these into two categories: general modeling and group modeling. The general modeling category will not be developed here, but involves such things as keeping the model simple, making sure it can be understood by the intended audience, ensuring that units within the model are correct, and performing sensitivity analyses. Of more interest here are the group modeling requirements related to submodel design and development.

First, the submodel must be developed within the parameters established for the project:

1. Name the variables and stocks that will be visible to other submodels in the final model in a manner that ensures uniqueness. For example, two subgroups may both be tempted to use the variable name "age." While this will work within each submodel, it will cause difficulties during submodel integration.
2. Use only available software and hardware.
3. Rely only on available external submodel outputs.
4. Generate all outputs required from the model.
5. Use and generate all inputs and outputs with respect to the units agreed upon at the group level.
6. Complete submodel development within the negotiated time frames.
7. Communicate all required changes quickly and with sensitivity to other submodel teams.
8. Continually monitor the submodel's internal state and external input variables to determine whether the submodel is operating within reasonable or experimental parameters.

Test and initially refine the submodels using the group-generated artificial time-series data. As the individual submodels are completed. integrate them into the full model.

Step 5—Develop Full Model

As submodels are completed as individual, stand-alone entities, they become available for initial integration efforts. Because each submodel was developed with respect to a common time-series output test environment, they should tie together in a seamless fashion with little or no conflict. Submodel integrators must, however, be careful to detect such problems as unit mismatch, two or more submodels using the same variable name (and yet meaning two different things), and unused or unavailable information.

Once the submodels are integrated and operating simultaneously, the submodels will be responding to input combinations different from the test environment. This will likely cause unexpected output behavior of the full model. To reduce the effort that will be required to isolate the source of such problems, submodels should include instructions for evaluating whether the submodel inputs are within acceptable ranges. An indicator can be then set to indicate that the submodel is working with inputs that are out of range.

Step 6—Iteratively Test and Debug

Generally, the full model will not initially behave as intended. Modeling teams will need to work closely together to evaluate the model as a whole and to monitor the submodel behaviors when operated within the context of the whole system. Unexpected cycling, chaotic activity, operation of submodels outside of their range of sensitivity, and just plain nonsense must be detected and analyzed. Required submodel changes are communicated to the appropriate submodel teams.

For the purposes of debugging, full model integration should proceed one submodel at a time; that is, two submodels should be tied together and tested. Once they are performing as expected, a third (or another set of two that have been tested together) are added to the model. Problems are isolated through full team efforts that involve careful inspection by each subgroup of the behavior of the submodels during full model integration and testing.

Once the development team is satisfied with the operation of the model, the end users are brought in for demonstrations and feedback. Time availability, perceived quality and flexibility, changes in user needs, and funding availability will affect further refinement or redevelopment opportunities.

8.2 Management Concerns

Interdisciplinary efforts have special management needs. Individuals from different disciplinary backgrounds often bring different paradigms to their problem solving. These differences set the stage for difficult communication, which can lead toward significant educational opportunities. However, here we want to focus on the management considerations that are directly involved with the modeling effort. Identified below are a number of issues that must be recognized by the individual or group that takes responsibility for the integration of the individual modeling efforts.

Submodel Ownership Issues

Large, complex, multidisciplinary modeling efforts undertaken by teams of individuals must be split into logical smaller efforts. The role that each smaller effort plays in the project is defined within the scope of the entire project. Each of the resulting submodels might be completed as the result of individual effort, imagination, and expertise. Development of the submodel begins with the definition of submodel requirements, continues with development generally accomplished in isolation, and is completed with refinement efforts during full model integration. It is crucial that a single owner of each submodel be involved at all phases. The temptation to "turn over" a model component must be avoided until the project is completed. If a component that has been "completed" early in the project turns out to need refinement or other modification, it is almost always more efficient for the author of that submodel to make the changes. In cases where another person is assigned the modification task, it can be often be easier to "start from scratch" than to take the required time to understand the original. For the sake of time, expense, pride in accomplishments, and avoidance of disruption, submodel ownership must be maintained throughout the entire project.

Submodel Development: A Part, Not an End

Submodels must be developed as part of the intended end product as opposed to an end in themselves. Developers must ensure that the submodels they are developing provide outputs required by the full model. On the other hand, the development of subcomponents that do significantly more than intended can waste development resources and cause simulations to run slower than might be tolerable.

What Holds the Group Together?

Interdisciplinary groups can be especially difficult to coordinate. Someone in this group must take the responsibility for pulling together the individual backgrounds, motivations, and expertise. When there are differences of opinion, someone must provide the leadership to make the difficult decisions. Select the best individual, for this position is probably the most important decision in the entire modeling process. Ideally, this individual must work well with all members of the team at a personal level and at a technical level. The person must be trusted to make good, carefully considered decisions, and have a very good understanding of the process by which the team will accomplish its goals. Choosing the right group leader is crucial to success.

Scheduling

People operate best within the context of achievable expectations. Schedules define these expectations within the framework of available time, energy, and ability. A team must work from a common schedule that provides reachable goals within realistic time frames. A realistic, yet productive schedule must be developed with and communicated to the team. Plan the modeling and model within the plan.

Feedback Requirements

The modeling process described in this book was divided into three phases: (1) overall model conceptualization, (2) submodel design and development, and (3) full model integration. These phases have somewhat artificial boundaries, as the activities in each are accomplished with respect to the activities immediately before and after. Hence, full model conceptualization is accomplished with feedback from the submodel design effort, which is completed only with feedback from the full model implementation phase. Indeed, the three steps define an iterative process where each succeeding full model builds on the successes and lessons from the previous effort.

Feedback from submodel design and development to overall model conceptualization may involve the following questions:

- Can identified submodels be developed with available expertise?
- Do the data exist to support conceptual model requirements?
- How long will the full simulation model run on available hardware?

Feedback from full model integration to submodel design includes the following questions:

- How do the submodels behave in the context of the full model?
- Do the chosen time steps capture the dynamics within the operation of the full model?

8.3 Conclusions

This section outlined a framework within which an interdisciplinary team of researchers can design and develop a large dynamic, spatial, ecological watershed simulation model. A balance was sought between efforts that involve approval from the team membership and efforts that require individual imagination, expertise, and motivation. Full model conceptualization is performed with respect to the end user requirements as tempered by available resources. Partitioning the model into portions that can be developed through individual efforts follows the conceptualization of the full model. As the model components are completed within the overall design requirements, the components are pulled together into a working final model. Full model debugging requires modification to the components by the original developers of those components.

Leadership in this multidisciplinary approach is crucial and must be approached with sensitivity to the personality, background, motivational, and time availability differences of the team members. Using the approach outlined here will assist the leadership as well as the team members in the successful design, development, and operation of large, complex spatial models.

9
Analyzing Alternatives

Previous chapters dealt with foundations for and approaches to modeling. They reviewed numerous modeling environments and models. Assume now that a model has been created and/or adopted to assist in a watershed management decision. Effective natural resource management involves the comparison of alternative management plans and scenarios. Opinions of local citizens and scientists, outputs of discipline-specific models, and results from large multidisciplinary models all predict consequences of the alternative plans. Deciding among options is often very difficult for a number of reasons. First, when multiple stakeholders are involved, different value systems can result in different rankings of the alternatives. Second, it is unlikely that a number of objectives can be fully met by any one of the management options. Objectives are typically competing and it is important for individuals to understand the relationships between a set of important objectives. In this chapter we explore methods to help choose among a set of options with respect to their abilities to meet a set of stated objectives. We then briefly look at the possibility of using the computer model to generate improved alternatives.

9.1 Decision Trade-Off Analysis

Our lives are filled with endless decisions. We associate decision making with freedom, pain, agony, opportunity, and self-improvement. We use a variety of approaches to make our decisions. Perhaps the most fun, but also the most dangerous, approach is the impulsive approach. Decisions are made quickly and without deep reflection. This works well for decisions associated with very limited consequences. A second approach is the "trust yourself" approach. Consider the alternatives through thought, prayer, and/or meditation and feel the value of each. This is perhaps our most common approach to making significant decisions. We encourage one another to "trust our hearts." Considering different scenarios can actually be associated with palpable feelings of disgust, peace, nausea, euphoria, or fulfillment. This approach taps into our complex internal conceptual models of the world that develop through our experiences in and with the world. This approach relies on sufficient accurate and authentic experiences that we can trust. Age is associated with wisdom, for through the experiences of life our internal models of the world are continually refined. Major decisions in business, government, and individual lives are almost always made in this manner. A major drawback to this approach is that language is not entirely adequate for communicating the reasons for making a particular decision, and in multistakeholder situations, communication is critical. Communication can occur indirectly through shared common experiences. This is why it is important for a watershed group to visit the watershed problem areas, see the faces and hear the stories of people affected by current problems, see the result of floods, smell the backwaters, touch the plants, and feel the strength and beauty of the natural systems, and tour businesses and people that will be affected. Even after the common experiences, however, groups will still have challenges in making collective decisions. A more formal approach to decision trade-off analysis can be useful in these situations. Simulation models can help with communication challenges in a number of ways. First, the very act of putting a model or set of models together can afford an opportunity to discuss the goals and objectives, study the available resources, and discuss individual and group concepts about the watershed system state and dynamics. Second, if a group can agree on, or accept, model inputs (state and dynamics information), it should be easier to accept model outputs and predictions. Watershed management often involves the art of making management decisions in situations of multiple stakeholders, multiple objectives, and multiple management strategy alternatives. Exactly where and how can spatially explicit simulation modeling be useful in this process?

Group decisions can be approached in a number of ways. A common approach is to simply move forward on decisions to which everyone agrees. We like the ring of unanimity, to leave a meeting and declare success without opposition. Unfortunately, such decisions can be rather hollow, without serious consequence, and involve "further study." But, after months or years of planning, these types of decisions can be the final culmination of a group effort. A modified version of this approach involves agreements between individuals to support decisions in exchange for having their own desires supported. For example, an "environmentalist" might support the development of a golf course in exchange for commitments from the economic development council to support the creation of a wildlife preserve. Such agreements can lead to a more comprehensive watershed management plan. Strong and effective leadership can be crucial in developing coalitions of individuals and groups that might otherwise be antagonistic.

Formalized decision processes for trade-off analysis have been developed to assist in decision making. To develop the ideas and approaches, imagine the decision to buy a car. A number of scenarios have been identified; that is, a number of different cars have caught your attention and you want to make a rational decision. For each car, you know the price, color, age, customer satisfaction, repair needs, miles per gallon, and horsepower. Although it may be too much to compare all of the cars at one time, you might be comfortable comparing the various qualities. For example, you might prefer higher horsepower and higher miles per gallon. As these are typically inversely related, you might be able to identify how much horsepower you'd be willing to sacrifice for an improvement in mileage of one mile per gallon. Similarly you could contrast age and customer satisfaction. After expressing your preferences at this level of detail, logic and simple algorithms can be applied to check for consistency in your answers and then identify the scenario (the car) that best addresses your preferences. A number of different approaches have been developed to collect information from a decisionmaker and then apply that information to the decision at hand. These approaches are available in commercial software.

9.2 Approach to Analysis

Figure 9.1 suggests a rather simple set of objectives and a matching set of actions. In reality, the objective list can be kept rather small, but the action

Multi-Objective Trade-Off Analysis		Objectives				
		Cost	Improve Business	Improve Biodiversity	Improve Community	Improve Water Quality
Actions	Host Watershed Fair	Hi	Med	Med	Hi	Lo
	Introduce Legislation	Hi	Lo	Lo	Lo	Med
	Stabilize Riverbanks	Lo	Lo	Med	Lo	Hi
	Buy Riverfront Property	Lo	Lo	Med	Med	Hi
	Re-meander streams	Lo	Lo	Hi	Med	Hi
	Advertise for Tourism	Med	Hi	Lo	Lo	Lo
	Study the problem	Lo	Lo	Lo	Lo	Lo

FIGURE 9.1. Multi objective trade-off analysis

list can be lengthy through creative idea generation. A polarized group might not find full consensus on any action except to conduct further studies. Several formalized decision processes have been developed to provide good, if not optimal, decisions in a complex decision-making environment based on relatively simple information extracted from the decision-making community. One approach is the analytic hierarchy process (AHP) developed by Saaty (1987). This approach has been captured in easy-to-use, decisionmaker-oriented software. One example is Expert Choice.[1] The AHP decomposes options or scenarios into components and those components into subcomponents. Decisionmakers identify trade-off interests at the lowest level and algorithms are applied to the human input to help identify trade-offs at successively higher levels of aggregation until a decision is identified at the highest level. Another approach is the multi attribute utility theory (MAUT) (Posavac and Carey 1989). These and other techniques elicit simple trade-off decisions from an individual or group that involve only part of the decision space at any one time. As the information required is collected, simple software calculations reflect back the consistency of the input. Once the user is satisfied with the responses, simple algorithms rank-order the suggested actions based on their perceived impact on objectives and the preferences of the individual or group in meeting the objectives.

A number of mathematical approaches can be used to combine the perceived impacts and objective preferences. Figure 9.2 is based on Figure 9.1, but replaces the Lo, Med, Hi rankings with numbers 1, 2, and 3, respectively. (Note that Hi in the Cost column indicates "Hi impact on a low-cost objective," not "Hi cost.") In addition, each objective is associated with a value to compare its importance with other objectives. A rating is simply calculated for each action by multiplying the objective importance with the associated action impact on the objective and then summing. For example, the action rating for the "Host Watershed Fair" action is:

$$3*3 + 2*2 + 2*2 + 3*3 + 5*1 = 31 \tag{9.1}$$

It is interesting to consider that perhaps the only action that a disparate committee comprised of environmental, economic, and community health advocates might agree on is to "Study the Problem." But in this sample analysis, anything is rated higher. Such a committee might find it challenging to identify the potential impacts of each proposed action on the stated objectives, but technical committees can be very useful in this matter. More challenging, however, is the development of group consensus on the relative importance of the different objectives. The group can experiment rapidly with various combinations of importance values and thus test the sensitivity of the simple model to the values. This exercise might result in a

---

[1] Expert Choice—http://www.expertchoice.com.

Multi-Objective Trade-Off Analysis	Objectives					Action Rating
	Cost	Improve Business	Improve Biodiversity	Improve Community	Improve Water Quality	
Importance	3	2	2	3	5	
Host Watershed Fair	3	2	2	3	1	**31**
Introduce Legislation	3	1	1	1	2	**26**
Stabilize Riverbanks	1	1	2	1	3	**27**
Buy Riverfront Property	1	1	2	2	3	**30**
Re-meander streams	1	1	3	2	3	**32**
Advertise for Tourism	2	3	1	1	1	**21**
Study the problem	1	1	1	1	1	**15**

FIGURE 9.2. Rate actions

number of proposed actions being clearly inferior; they can be removed from further analysis.

More rigorous approaches are available for identifying the importance values (e.g., as captured in the Expert Choice software). For example, the law of diminishing returns indicates that the value of a marginal increase of something drops as one has more of that thing. For example, a slight improvement in the quality of water is much more important when that water has poor quality than when it already has good quality. The value of an acre of land for supporting biodiversity makes a big difference if it is the first acre, but a much smaller difference when it is the thousandth acre. Consider the objectives in Figure 9.2. In our simple action comparison we have indicated only one number for capturing the relative importance of each objective. This can be sufficient if we are making only marginal differences in meeting objectives, but, as those objectives are met, the importance of further meeting the objectives typically decreases. An environmentalist will be happy to support economic growth once the environmental objectives have been "adequately" addressed. Similarly, local business interests will support environmental objectives once the business needs have been addressed sufficiently. A more complex approach is required to define the trade-off curves, and this is where the power of procedures like AHP and MAUT becomes important.

9.3 Summary

Watershed management groups seek to simultaneously address a number of objectives by allocating resources to a number of proposed actions and activities. Modeling is not useful for identifying goals and objectives or in proposing alternatives. The role for watershed models is in the prediction of anticipated consequences associated with proposed actions. Watershed modeling is most profitably employed when this context is fully understood. Once models are used to carefully connect management alternatives with objectives, alternative trade-off analyses reviewed in this chapter can be useful.

10
Who Develops and Runs the Models?

This chapter considers three alternative approaches to the running of simulation modeling for land and water management decision making. The scientific community is ultimately responsible for the definition of many of the dynamics that drive simulation models. Hence, the first approach is to have the scientists run and interpret the models. The second is a two-tiered simulation modeling approach in which scientific models are used to set variables in simpler management models. The third approach is to create management models from scientific models.

Before reviewing these alternatives, it is important to understand why scientific models are generally inappropriate for management applications without significant conversion. What are the differences between "management" and "scientific" models that require a conversion effort? Scientists typically make progress by isolating some component of a system under consideration. Isolation is accomplished by holding constant as many system variables as possible. Knowledge of the system's state and dynamics is teased out through experimentation based on the scientific method. The requirement for isolation of system components for study supports the structure of the scientific community that we have today: a set of disciplines that are frequently isolated. Each discipline publishes new insights of nature through studies of the behavior of "controlled" systems. These insights are described as models of the system and the models can be captured in computer software as spreadsheets, statistical correlations, dynamic simulations, series of equations, and assorted algorithms. Because the models, by the nature of the scientific process, are focused on small components of larger systems, they are rarely useful to the manager of natural systems. Though they may have proven accuracy under laboratory conditions in which most of the system is tightly controlled, they are rarely useful in natural settings where none of the dynamics are controlled.

A second challenge to land and watershed managers (that is not a concern to scientists) is defining the goals and issues for the management process to address. Management can rapidly become extremely difficult and complicated. Rittel and Webber (1973) used the term "wicked" to characterize a class of problems. "Tame" problems are those that can be handled easily with currently known information and developed technology. They can be very complex problems, but they have already yielded to human efforts to understand and control them. Rittle and Webber identified the following set of properties that define "wicked" problems. Watershed managers will easily recognize their challenges.

- The problem is difficult to define. Stakeholders are easily distracted into arguments about definitions of the problem to be solved. Stakeholders find it difficult to mutually identify the problem.
- The problem can be approached at various levels of resolution and abstraction. Complex judgments are required to determine the appropriate level to tackle.
- It is impossible up front to identify the conditions under which a solution fully succeeds.
- At best, solutions can only be judged as better or worse and never fully right or wrong.
- The success of any solution cannot be judged objectively.
- Alternative solutions are discovered in the process of application of a solution.
- Strong moral, political, or professional components often play a key role in seeking solutions.

Watershed "wicked" problems can be associated with strong multiple stakeholder interests and/or a high degree of scientific uncertainty (refer to Fig. 2.4 in Chapter 2). Noting that the watershed manager is faced primarily with "wicked" problems is not admitting defeat or suggesting that science, technology, and spatial simulation modeling be ignored out of hand. It does, however, suggest that how such models are used must be tailored to each individual circumstance. The following discussion covers three distinct approaches to using dynamic simulation modeling to affect decision making.

Assume that your watershed management group or county land management office has decided to employ a dynamic watershed simulation modeling approach to the evaluation of a series of alternative land management options. Perhaps they are working toward a decision with respect to the impacts of the options on local wildlife and storm-water flow and flooding. You are now considering how the group should proceed with the modeling effort. Three basic choices should be evaluated with respect to your available time: funding, accuracy needs, and talent. The first option discussed below is to turn the assignment over to the scientific and engineering community. You collaborate with a set of experts who study the options with their in-house models and expertise and you receive the results and interpretations of the simulations at a later date. In the second approach, you adopt a two-stage modeling approach. Management models operable in a management office get their inputs from results generated by complex

FIGURE 10.1. Scientists use models

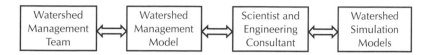

FIGURE 10.2. Management uses a management model

scientific and engineering models operated by the scientific experts. For the third approach, local management models that have the complex scientific and engineering models embedded inside are run. Each approach has its own set of costs and benefits that are presented below.

The first approach (Fig. 10.1) is to hire a scientific consulting team or individual to conduct the modeling, interpret the results, and provide feedback in the form of a report. This is the traditional approach, typically accomplished through a contract to a consulting firm, a university, or a government agency. The experts, at their discretion, use available simulation models fed with information and data gleaned from various sources. The benefits of this approach are that the local decisionmakers need not have any particular expertise in the modeling process. The role of science in the decision can be left completely in the hands of the "experts." Another benefit is that the experts can make use of cutting ("bleeding") edge technologies that are not ready for general use. The drawbacks can be significant. Upon completion of the simulation, the product is nothing more than the expert's report. Although this may be sufficient, it can be expensive to update initial inputs or change the analysis in any way. It is also difficult to check the work that was done. This approach requires a full measure of trust between the client and the consultant.

The second option (Fig. 10.2) is to use a watershed management model that has been developed specifically to address the local management concerns. This model can be anything from a relatively simple information system to a complex and involved decision support system (DSS) developed to rapidly assess management alternatives. Information systems can take the form of published books, data and information compiled on CD-ROM, or Internet sites containing data, instructions, and ideas. Government agencies and universities continue to compile vast amounts of information into Internet-based information nodes. As information technologies improve, we are seeing access to more information useful at the scale of a local watershed.

FIGURE 10.3. Management models linked directly to scientific models

The third approach (Fig. 10.3) is to embed scientific models behind management decision models. The idea is that a decision-making body might brainstorm a number of alternative land management schemes. Instead of working indirectly with simulation models through scientists and engineers or indirectly through simple management models parameterized by scientists with their scientific models, the land managers run management simulation models that have embedded scientific models. This approach is not yet common due to complexities in the integration and maintenance of disparate simulation models. Government agencies and universities are currently working to provide access to large management-oriented simulation models that rely on behind-the-scenes linked scientific models. Instead of packaging these models for operation on local end-user computers, Internet-based user interfaces are being developed that will execute models over the Internet at run-time using approaches discussed later in Chapter 13.

These three approaches reflect a progression from the traditional past, through the present, and into the future. The principal methods of delivering better decision support to watershed decisionmakers is through publications, Internet databases and information sites, and simple management-oriented decision trade-off analyses that are based on the knowledge gained from scientific investigations and complex scientific models.

11
Error and Uncertainty Analysis

In our private lives, we make decisions that affect our personal future and our community's future based on conceptual models. Accurate models that result in predictable futures bring success and satisfaction to our lives. Each of us is blessed with some level of capability to absorb the lessons of the past into our personal models that then help us move toward a desirable future. We are all aware of the difficulties associated with making decisions when the model is incomplete—either because we lack confidence in the model or we are seeking additional information as inputs to the decision process. Sometimes we accept the fuzziness in our thinking and the available information and make a decision with full awareness that the decision will need to be revisited as our understandings (our models) improve and input information is made more complete. Sometimes we are more comfortable with "black and white" and find ourselves turning a "gray" area decision into a confident yes or no. Uncertainty is uncomfortable for most of us, and we can find it easy to move to adopting ideas or decisions that appear to have more certainty than is actually the case.

Socrates exposed the folly of the human penchant for making choices based on our desire to avoid uncertainty. In his time, the most powerful individuals were those who could, through force of personality and power of oratory, move an audience or a population to accepting an argument. Those people knew well how to use the human desire for certainty and the human ability to associate strong emotions with importance. People could, and of course still can, be swayed with simple passionate oratory. Today, tent revivals, charismatic leaders, and political rallies provide opportunities for us to witness and participate in this ancient tradition. Socrates, using simple and easily accepted formal logic, was able to show audiences that the conclusions drawn by the great orators sometimes did not make sense. This was so unsettling to those in power that he was put to death. The fundamental human nature that allowed humans to be swayed by well-communicated passion still exists in our nature and those who believe in the supremacy of formal logic must still fight Socrates' battles every generation.

Personal conceptual models based on thoughts and feelings born of unspecified mental processes that seek to find pattern and consistency in life must deal with error and uncertainty in the model and the model inputs. We are very limited in our ability to deal efficiently or effectively with error and uncertainty. In many cases, we prefer to not deal with conceptual models at all—we prefer instead to simply accept conclusions. Here is a simple experiment to test this hypothesis. When someone comes to you for advice, begin by flipping a coin. If heads, give a full answer that delves into various arguments for and against each factor so that your questioner has more information that they can use to come to the best answer. If tails, provide a concise answer and deliver it with a strong emotional sense of authority. For most people, which answer will be most satisfying? Most of us, most of the time, want straightforward, unambiguous answers. With such answers, such simple models, there is no opportunity to evaluate potential sources of error or uncertainty. Only by using Socrates' approach to the application of formal logic do we develop our best opportunity to evaluate errors and uncertainties. Because our human minds struggle with the application of logic in large models, it becomes important to formalize larger models by writing them down. Unfortunately, it is very difficult to communicate a personal conceptual model to another through language. Conceptual models are best duplicated through a sharing of common experiences. Then, only through developing associated formal models in a common language can we collectively begin to evaluate the impact of model errors and input uncertainties.

11.1 Sources of Error

Any model, conceptual or formal, is associated with errors and uncertainties. The sources of these errors are many and can leave one feeling very uncomfortable and uncertain. Here is a brief review of error sources.

Errors of Exclusion

When developing a model, it is important to explicitly consider all parts of the system being modeled. Because it is impossible or impractical to model everything, the modeler must ignore those components of the system that are least likely to have a significant impact on the model output and on the targeted management decision. This process of informal sensitivity analysis is critical, but can result in the removal of a key aspect of the intended model. For example, a watershed model focused on economic growth might not capture the impact of water quality issues on the ability to attract a high-quality workforce. The resulting model might suggest that businesses with lower environmental restrictions will grow, when in fact, that potential growth might not be predicted if this effect had been part of the model.

Errors of Inclusion

Errors of inclusion have a more subtle and indirect effect. Because it costs time and effort to develop any part of a model, developing a model component that will not be critical to scenario evaluation decreases the quality of the more important parts of the model.

Incorrect Algorithms

Ideas about the cause–effect relationships in a model can be inaccurate, inappropriate, or misleading. Very involved, complex, and inaccessible model components might be very impressive to the layperson, but if incorrect, those components can be very damaging to the model's ability to estimate future consequences.

Inappropriate Interpolation and Extrapolation

Scientific facts and discoveries drive models. In most cases, the experimental design limits data collection to well-defined and well-controlled conditions, which allows cause–effect relationships to sometimes be established. What is really known then, is that this cause–effect relationship holds true under the experimental conditions. It is then very easy for a modeler to adopt this relationship. Within the model, environmental conditions will change and interpolation and extrapolation of the experimental data can occur. Extrapolation is particularly difficult to justify, though it is easily and invisibly accomplished. For example, a researcher may establish a relationship between ambient temperature and food intake for a given species. Extrapolation of the discovered relationship will break down when the temperature gets too hot or too cold. More insidious is the extrapolation of the relationship to different environmental conditions (e.g., different water or food availability, or different levels of competition for food) or the extrapolation to different populations of the same species that may have developed a slightly different food–temperature relationship.

Inappropriate Time and Space Resolution

Traditionally, many environmental models were developed as analytic models, formed with sets of simultaneous differential equations. While they are still powerful modeling approaches, analytic models require that the equations be continuous (no environmental thresholds) and they tend to be rather inaccessible to anyone without many semesters of calculus training. A less accurate, but more flexible and accessible, approach that is quite amenable to solution with modern digital computers uses difference equations. Time and space are divided into well-defined steps, and relatively accessible algebraic and logic equations are developed to step a simulation model from one time step to the next. The equations, when applied over and over, will move the system state through time. Information about the watershed is stored for each plot of land or water at a regular predefined spacing. The difference equations are applied simultaneously to each of these plots and can base the state of the plot at a next step on its and its neighbors' previous states. The choice of resolution can be critical, and the time and space resolutions interdependent. Interdependence occurs when some quantity (e.g., water) is moving across space. If that movement in a given time step is further than the resolution of the space, difficulties will occur. If animal territories are being modeled, it is easiest to set the spatial resolution to approximate the size of these territories. Modeling territories as collections of sites is much more challenging and if ignored will be inaccurate.

Inappropriate Time Step Algorithms

As described in the previous paragraph, difference equations are commonly used to move a watershed model through a series of future states. The easiest method for computing the state of a next time step is to directly apply the algebraic and logic-based formulas using the state of the previous time step. Unlike the application of smooth differential equations, this approach can cause values to change substantially in a linear fashion. Shortening a time step might result in a more accurate future scenario. Unfortunately, halving a time step requires a doubling of the computational time. Several techniques have been developed to get the effect of time-step shortening without the full increase in computational requirements.

Incorrect Inputs

Models are driven by an initial system state and a set of equation coefficients—each the result of a scientific analysis of field or remotely sensed data. The process of turning measurements of field information into accurate model inputs is fraught with the potential for errors. The error potential starts with the data collection design and continues through the collection and interpretation of those data. Ideally, an audit trail is available to trace each model input back to the collection of those data. In practice, this is expensive and unlikely.

Inappropriate Order of Execution

Difference equation–based simulation models must have their equations solved for each spatial location. Using modern digital computers, those equations are typically solved in some set order, but the order of execution can have a dramatic effect on the model output. For example, consider a population spreading across the watershed where death in that population is a function of the population density. If density-based death is computed before population spread, the population will be smaller than if spread is calculated first. This type of effect can result in dramatically different model output trajectories and is easy to overlook.

Inappropriate Conclusions

Our human penchant for using words like *always* and *never* and for accepting the pronouncements of experts can lead us astray in the conclusions that we draw from the model. If a model of an endangered species projects the demise of a local population, it is not appropriate to conclude that indeed that population is certainly doomed. All that can be said is that a specific model associated with a long list of assumptions suggests that the population might be in trouble.

Other sources of error exist, but this section has certainly made you aware that the sources of error in any modeling effort are substantial and sometimes surprising. The challenge of the modeler is to identify the actual errors associated with a model and understand the implications of those errors with respect to the management decisions under consideration for which the modeling was accomplished.

11.2 Tracking Error and Uncertainty

Although all measurements are associated with uncertainty and the choice of algorithms is critical to prediction accuracy, errors and uncertainties are often neglected as part of a data set. When knowledge of errors and uncertainty is missing, it is virtually impossible to evaluate their impact on simulation model output. The human ability to ignore the potential implications does not, unfortunately, decrease the importance and impact of the errors and uncertainty.

Consider the implications of errors on a simple model. Figure 11.1 depicts a STELLA model containing a single state variable called "Population." Deaths are calculated each time step as:

$$\text{Population} * \text{death\_rate} \tag{11.1}$$

The birth rate equation is:

$$\text{max\_br} * [1 - (\text{Population}/\text{max\_pop})] \tag{11.2}$$

The deaths and births equations are:

$$\text{Population} * \text{death\_rate} \tag{11.3}$$

$$\text{Population} * \text{birth\_rate} \tag{11.4}$$

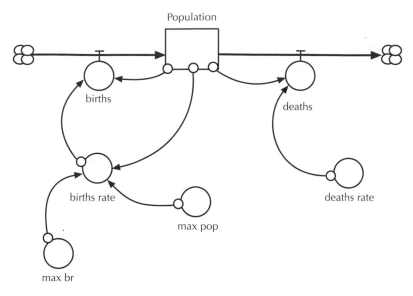

FIGURE 11.1. A simple population model

Variables are set as follows:

Population = 1000
max_pop   = 1000000
max_br   = 0.3
death_rate = 0.1

Errors can be associated with any of the variables, including the initial population, and can be associated with the choice of equations. In this case, the birth rate equation is the most questionable component. It is a simple logistic equation that decreases the birth rate smoothly as a function of population. Although the concept may be correct, the detail of the algorithm may be incorrect. Consider error or uncertainty in the initial starting population. If the population estimate is too low, then the births in the first time step are also too low, yielding an even lower prediction of the population in the second time step. This implication of an initial underestimation will result in an ever increasing prediction error. Similarly, overprediction of the initial population results in an overprediction of births, yielding increasingly overinflated populations in the future. Similar kinds of error propagation can be associated with the other variables. More complex uncertainties that are less tractable can be associated with the choice of algorithm.

There are a number of ways to track and evaluate the impact of model errors and uncertainties on model outputs. Monte Carlo simulation is a popular approach. Imagine that a model contains a single variable and that something is known about the range of potential values and the probability

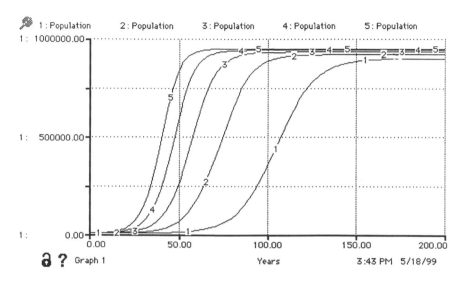

FIGURE 11.2. Initial population sensitivity analysis

of each. The model is run a number (perhaps many) times, each time using a randomly chosen value for the variable based on the known probability distribution. After many simulation runs, a distribution of the model output will emerge that communicates something about the sensitivity of the model to that input. Some simulation model development environments (e.g., STELLA) provide some atomation in the evaluation of the sensitivity of model output to variable estimation. Figure 11.2 shows the trajectory of the population (from the model in Fig. 11.1) using birth rate values of 0.1 (1), 0.2 (2), 0.3 (3), 0.4 (4), and 0.5 (5). After 50 years, the projected population varies between about 10,000 and about 800,000. By 150 years, the population difference is based primarily on the fixed death rate. Where many variables are involved, many runs are made using randomly chosen values for each of the variables, or a separate series of runs is made for each variable.

Sensitivity analyses are important for several reasons. First, they formally evaluate the range of potential futures allowed by the model and can give a probability distribution of those futures. Second, it can help focus the expenditure of limited resources on the collection of data by identifying those uncertain inputs most dramatically affecting the model's predictions.

12
Model Evaluation Guidelines

So far, watershed modeling has been presented in this book with respect to underlying approaches to hydrology and ecology, existing models developed out of different academic traditions, how to approach model creation and applications, and modeling costs and benefits. While many models and modeling environments have been listed and cursorily reviewed, they have not been formally reviewed. Evaluation must be based on the particular needs and constraints of the watershed management group. This chapter lays a foundation that a group can use to evaluate models and modeling environments with respect to local needs.

12.1 Requirements Identification

The first step in model evaluation is to establish a firm group understanding of the management questions. These questions will help the group identify potential modeling requirements.

Requirements for developing simulation models involve dynamics of human and group decision-making processes as well as the relevant dynamics of the natural and human systems playing upon the watershed. Let us first consider the human decision-making component and then delve into the more scientific requirements.

Before undertaking any effort that requires a significant commitment of time, effort, and financial resources, it is important to thoroughly identify the requirements. Remembering that the development of digital geographic information system (GIS) databases and the specification of land and watershed dynamics are information-formalizing processes, the requirements for information formalization must be evaluated. If a group of citizens collectively agree, there is no perceived requirement for expending any resources to support that agreement. In a watershed scenario, if all players in the management of the watershed believe that it is appropriate and necessary to ration fertilizers (or to make fertilizer even more accessible through local subsidies), then political agreement has been reached and expenditure of any resources to substantiate that agreement is wasteful. If, however, there are serious disagreements in the community, a requirement exists to further explore the issues and find a workable solution. Formalized watershed modeling can help in three fundamental ways. First, it provides a focus for communication among the parties. Such communication can result in agreements before any modeling is actually accomplished. Second, it forces participants to go beneath the emotion to support their beliefs with logic and facts. And third, it encourages the involvement of scientists who have deeper knowledge of the particulars of economics, ecology, hydrology, and other regional dynamics. If the group agrees to employ modeling, first identify the group decision-making requirements that can be served by modeling. What level of modeling is required to address these needs? Perhaps individual conceptual models of the land or watershed can, through dialogue and common experiences, be transformed into a conceptual model owned by the group and perhaps the community. Perhaps the consequences of alternative management options are well understood and accepted, but there is no best option due to value differences among individuals. For this situation a trade-off analysis might help a group find win–win opportunities in the options. Finally, a group might find that they need to reach out to the scientific community for unbiased, value-free analyses of alternative management strategies. Identification of the group decision-making requirements and opportunities is essential before a modeling effort is properly defined.

When a management group turns to scientific data and models, the salient aspects of the associated human and natural systems must be identified. It is useful to begin with a statement of the decision or decisions to be made. What are the local issues of contention? Can a group formulate a question or set of questions that, when answered, will help move the group toward consensus?

If land and watershed management consensus can be significantly improved through watershed simulation modeling, it then becomes necessary to define and focus a planned effort. A repetitive split–evaluate–focus scheme is useful for defining the modeling effort without forgetting any important components. Begin with an acceptance that all parts of the system under consideration are connected in a full ecosystem. Split this system into a few (e.g., three to seven) subsystems as appropriate for the defined questions. Perhaps the group might identify hydrology, weather, animals and plants, and economy as the subsystems. Ensure that the set of identified subsystems encompasses all of the important portions of the full system—again focusing on the established questions. Then, evaluate each of the subsystems with respect to their relative importance for addressing the questions. Remove those subsystems that appear to offer insufficient benefit for the perceived cost of further model development in that area. Repeat the split–evaluate–focus process with each of the subsystems to identify even finer detail. For example, a hydrology submodel might be split into drinking water quality, stream bank erosion, flooding, and water resources for crops and wildlife. Continue this process until the management group cannot easily accomplish further splitting. The model components that are left form the foundation of the model requirements. For example, a watershed management group might have settled on the following model components:

- Economic growth
- Attraction of the community to a new workforce
- Fish and mussel health
- Soil sediment loads
- Sheet and rill erosion
- Stream bank erosion
- Waterfowl populations

When challenged that perhaps nitrate levels in drinking water, farmer best-management practices, or flooding needed to be considered, the management group has a record of the process that started by considering everything and then narrowing down to connecting the modeling objectives to the short list of model components. Section 8.1 in Chapter 8 explored in more detail the process through which models are developed.

12.2 Establishing Modeling Expectations

Two extremes must be avoided when working with models; both are associated with unrealistic expectations. Models can easily be challenged as useless. Errors and uncertainties in modeling can be argued as fatal flaws. Models are always useful as conduits for formally stating how someone or some group believes the system works. The formality of this statement lays open any model for criticism and ridicule. Without the model to inspect, we are unfortunately reduced to arguing consequences of conceptual models that are not shared by the individuals involved. Arguing who is right and who is wrong necessitates a formalization of the logic and thought process by which we justify our viewpoints. By opening our thought processes, interpretations, and conceptual models of how the world works, we open ourselves to critically evaluating our own and others' ideas. The other extreme must also be avoided. The time and energy associated with the formalization of a model can so focus the modeler that it becomes easy to believe that the model is indeed a full representation of the real thing. The model can become more real than the real world. Models can never be complete. Two famous quotes should be remembered by all who build or apply models:

- George E. P. Box: "All models are wrong, but some are useful."
- Henry Theil: "Models are to be used but not to be believed."

What does model development cost? There are a couple of rules of thumb. First, modeling becomes more expensive as it becomes more detailed and realistic. Games such as SimCity[1] provide an excellent, though generic, way for one to experiment with trade-offs between objectives such as jobs, population, tax base, crime, transportation, and health through the management of the spatial arrangement of homes, buildings, parks, and businesses. Relationships among the parts of an operational city are defined in unseen and unknown algorithms that provide a good learning experience, but are inappropriate for real urban and regional planning. At the other extreme, a computer programmer must be hired to assemble, or perhaps write, a locally specific simulation model.

Second, modeling expenses increase as the salient features are less linked to the first principles of classical physics. The movement of water has been relatively well understood for centuries and modern hydrologic simulation models are typically based on principles captured in equations that were first defined a hundred or more years ago. Our understanding of the movement of water is based on two assumptions—laminar flow and steady state—that are never true in nature. Lots of empirical data and constants derived from empirical studies complete the equations. Hydrologic models abound and can require input data only for the area of concern. Modeling ecological processes for small areas requires the development of not only model-input data, but of the model itself. Chapter 7 discussed the process of using modeling environments for addressing this particular need.

Third, the more extensive the model, the more expensive. Scientific models that focus on a small component of a dynamic system can be ready to apply to a watershed management question—if that question is similarly focused. Watershed management typically involves finding decisions that balance between competing objectives; scientific models are completely inadequate. To support management decisions it is necessary to combine relevant scientific models; this is typically very expensive.

In summary, a watershed management group must establish appropriate expectations for the role of modeling in supporting the group dynamics, and it must understand the modeling costs and real benefits.

12.3 Model Selection Criteria

Let us assume that you have decided to seriously consider using simulation modeling to help with a land or watershed management decision. You have asked staff members to identify potentially useful models and modeling environments. Internet searches, phone calls to software distributors, visits with government and university researchers, and visits with colleagues have

---

[1] SimCity, developed by Maxis—http://www.simcity.com/home.shtml.

turned up a number of possibilities. How should you now evaluate the identified options? A number of key questions are presented below, each followed by explanations.

What Decision Is the Model Intended to Support?

"Don"t give me any information that I'm not going to use in a decision'— Emmet Grey, Chief of the Environmental Office, Fort Hood, TX. Land and watershed managers understand the importance of this statement. Their jobs are all about decisions and dealing with data, models, offices, and individuals. Information that is not useful in decision making is an inefficient use of time. Useful information can either be immediately useful, or add to background information for making future decisions. Most watershed simulation models currently available remain focused within a single discipline. As such, the decisions associated with the models typically consider management impacts on hydrology, ecology, economics, health, or aesthetics. Identify the new information that the model is intended to generate and determine if that information is valuable for decision making.

Is the Model Appropriate for the Region?

Ecological models, in particular, tend to be developed for applications in a particular region. Hydrologic and economic models are most often insensitive to regions. Make sure up front that the model under consideration is applicable to the particular area and requirements.

What Is the Current Status of the Model?

Is it experimental, public domain, or fully commercial? Who has used it successfully? When was the last version released? What is required to fix bugs and extend the model's capabilities? Generally, the more useful models are under continual refinement and have a user community, and user support is available.

What Information Is Required as Inputs?

To generate useful output, the model must apply its mathematical and logical algorithms to inputs. The inputs generally take the form of current (starting) conditions of the area and, for ecological models, involve locally specific behavior rules. Exactly what information is needed, in what format, and (easily forgotten) in what units? What will it cost to develop, reformat, and/or assemble the required information? How long will this take and who is available with the appropriate expertise to develop the inputs?

What Outputs Are Available?

Your application of the model is to generate information that will be used in a decision. What information will the model generate? Is it appropriate to the decision and in a form that can be useful in a management setting?

What Capabilities Exist for Import of Information from Different Sources and Formats?

Most modeling environments have data import capabilities that minimize the efforts required to prepare inputs. For spatially explicit input data, it is important to identify the capacity of the system to reformat, reproject, interpolate, and extrapolate data. What capabilities are available for importing data from statistical and spreadsheet data formats?

What Are the User Requirements for the Model?

What background and skills must the user bring to the model for success? Does the user community have those skills? Can the software be used only occasionally with success, or does a user need to work to retain the necessary skills to operate the system? Reviewing the model user interface can often provide answers to these questions. The more sophisticated the user interface, the shorter the learning curve, the better the on-line support, and the fewer technical skills required.

What Kind of User is Targeted?

A land or watershed manager responsible for management decisions needs to have a very user-friendly software environment that can readily reflect the consequences of alternative decisions, suggest alternative decisions, and help balance among competing multiple objectives. Few models provide this level of sophistication, but the list is growing. Even this type of interface requires a behind-the-scenes technical effort to prepare the inputs and assemble the model components. Most common are systems that must be directly operated by the technical staff in response to management requests for information.

What Are the Hardware and Software Requirements?

Currently available simulation models have widely varying hardware and software needs. Some well-developed models have been available for many years and run under DOS. Others are so central processing unit (CPU)-intensive that sophisticated supercomputers or workstation networks are required. Most of the models currently under active development are compatible with Microsoft Windows and few require more CPU horsepower or hard disk space than what is commonly available on current personal computers.

What Does It Cost to Use the Software?

The initial cost and the costs for maintenance, training, and support should be determined.

What Are the Time Requirements?

How long will it take to set up the system, prepare inputs, learn the system, and execute runs?

What Educational Material Is Available?

Are on-line material, text, articles, and source code documentation available? Are technical support and training available and at what cost?

Is the System Designed to Be Predictive?

Can predictions be easily compared, ranked, and contrasted? Does the system suggest alternative inputs or initial conditions to better meet user interests?

Does the Software Support Collaboration?

Watershed management can require significant interactions among managers, stakeholders, scientists, engineers, and politicians. Software that supports collaboration among a number of parties will yield results that the various participants can more easily accept. Also, some model development requires teams of interdisciplinary technical people. Does the software support different individuals or teams developing different parts of the same model? Can the participants be geographically separated?

What Kind of Technologies Does the Software Use?

Software technologies include deterministic process modeling, empirical modeling, fuzzy logic, inductive reasoning, knowledge-based reasoning, optimization, simulation technology, stochastic process modeling, symbolic logic, and others. Are the technologies acceptable to the decision-making community?

For Hydrologic Models, Which Components of Water Movement Are Supported?

Hydrologic modeling is a very mature field and many different approaches can be found in available models. Are watersheds divided into regular cells, triangulated irregular networks, or simply subwatersheds? Does the model support overland water flow, groundwater movement, and stream and river flow? Is sediment transport and deposition considered? Is energy conserved and momentum considered? Does the model consider water chemistry, especially nitrogen, herbicides, and pesticides?

What Kind of Modeling Is Supported?

Depending on the user and the stage of model development, the management group may need to do conceptual sketchpad modeling or may need to create highly formalized scientific models. Future modeling environments are expected to seamlessly handle a variety of modeling approaches, but today's capabilities generally focus more on the formal modeling.

How Are Errors Handled?

A major challenge to transitioning scientific models to management-oriented models is the way software errors are handled. Scientists can better tolerate terse error messages and unexpected software termination. Sources of errors are numerous and it is easy for the people involved with a management plan to ignore a model based on known input or algorithmic errors. It is important for any model to be able to calculate confidence bands on outputs based on an analysis of the impact of error and uncertainty in the model itself. This is rarely available in current simulation modeling environments.

How Are Sensitivity Analyses Handled?

Model sensitivity analyses are extremely important and should be available to the watershed manager. Unfortunately, many systems do not provide any automated ways to perform these analyses and sometimes models are too large to allow adequate analyses to be completed with available computing power. At the very least, the model should be able to run multiple times, each time with slight modifications in input variables to test the sensitivity of outputs to changes in those variables.

For Spatially Explicit Modeling, What Fundamental Data Type Is Used?

Spatially explicit data are captured in a number of different fundamental formats. Each type of data is matched with a wide variety of detailed formats—some public and some private. Examples of fundamental formats are raster, vector, hexagon, TIN, point, line, networks, and so forth. Conversion of data among different vendors' raster formats can be very challenging and expensive. Conversion among different fundamental data adds another level of complexity, uncertainty, and source of errors.

What Are the Spatial and Temporal Resolutions of Inputs and Outputs?

Time and space scales of inputs and outputs are critical. Check that the requirements match what you might have available in data sources.

What Ecological Scales Are Used?

Ecologists focus their studies at a variety of different organizational scales. Examples include individual, species, community, and ecosystem scales. If you are interested in projecting the biodiversity of an area under different management practices, you might find a community succession or a species-based successional model useful. Input requirements will be dramatically different.

What Other Models Are Required to Support an Analysis?

Some models, especially public domain and university-based systems, are built upon one or more software environments that may include programs and/or software shared libraries. In some cases, there are software tie-ins with separate products that support spreadsheet analysis, statistical analysis, visualization, or Internet publishing. It is best to clearly understand what hidden capabilities must be acquired.

12.4 Modeling Environment Selection Criteria

Modeling environments are most useful for modeling efforts that cannot be accomplished with preexisting models. People making land management decisions based on hydrologic concerns will most likely be overwhelmed with the hydrologic modeling options. People with biodiversity, community succession, and/or habitat suitability interests will likely not find useful ready-to-use management-oriented dynamic simulation models. For these applications we can turn to a software engineer or a modeling environment. Modeling environments capture a large amount of the software that is part of a final simulation model. Ideally, the only thing missing is the set of equations that describe the environmental system under consideration. Such things as a basic user interface, data input and output, model visualization, and network support may be part of the environment. "Modeling environment" is defined broadly here and includes everything from a basic software programming language to a full-featured easy-to-use graphical user interface that requires no computer programming skills. The considerations presented below are divided into three major categories: model development for programmers and model development for nonprogrammers. The first category considers a number of software engineering issues that must be considered when evaluating software that provides capabilities for software engineers. The second establishes considerations for evaluating software developed for nonprogrammers.

Model Development for Programmers

This section discusses software that helps software developers be more efficient in developing simulation models. Examples of such software include DIAS (DIAS 1995), SWARM (Hiebler 1994), MODSIM (Belanger et al. 1989), and HLA (DMSO 1996).

Language

Of primary importance to the software developer is the programming language associated with the modeling environment. Dominant languages today include VisualBASIC, FORTRAN, C, C++, and JAVA. Modeling environments may use other languages as well. Mixing languages is sometimes possible, but should not be taken for granted. Use of a language also points to a programming philosophy, which might be procedural, object-oriented, or even declarative.

Operating System

While many languages can be compiled for any operating system, software may be written to access some capability peculiar to a given operating system. In many cases, similar capabilities exist across a number of different operating systems and software can be written to make the appropriate system call depending on the operating system under which the program is compiled. More and more software is currently being written that communicates at run-time with other software. The approach to enabling this interprocess communication often varies among operating systems.

Libraries

Extensions to programming languages typically take the form of software libraries. Public domain and university-based software libraries are themselves dependent on other libraries. It is important to trace back through the software library requirements—including the required versions.

Development Interface

Special development interfaces are not typically associated with programming languages to support modeling and simulation. In some cases, a special interface speeds the design and development process.

Ability to Adopt Existing Models

Although model integration is important, redevelopment of models in an integrated fashion is often too expensive and breaks the connection with the original author. Some simulation model development environments provide software to facilitate the adoption of legacy (existing) software. Integration of existing models can take many forms, including compilation of models together into a single program, shared memory solutions allowing two or more programs to interact through shared main memory, and network interactions that allow two or more programs to communicate at run-time even if they are running on separate machines.

Object-Oriented, Declarative, Procedural?

Software languages can be roughly separated into these three categories. Most traditional software using languages like FORTRAN and C is procedural. Programmers specify the algorithms or procedures that are applied to data. Object-oriented languages are also generally procedural, but they encourage the development of self-contained objects that can reflect the way that we naturally perceive the world. Objects have a set of characteristics and respond in internally specified ways to requests from other objects. The object-oriented programming approach allows us to mirror these perceptions in software with some distinct advantages. Declarative languages are less well adapted, but provide a powerful alternative for certain software development exercises. Declarative languages allow the programmer to state or declare facts and relationships among facts. Using rules of formal logic, the declarative language allows the information to be queried for inferences that can be drawn from the facts.

Rule-Based Artificial Intelligence, Neural Nets, and Genetic Libraries

Some simulation modeling environments are associated with various libraries that support artificial life research. This line of research addresses, in part, theories of evolution and cognitive processes, which are supported by neural network and genetic algorithm capabilities. These libraries can be useful when modeling the behavior of higher vertebrates.

Interprocess Communications

Traditional software is written as stand-alone programs. Some emerging capabilities require that two or more separately running programs (and perhaps running on separate machines) communicate with one another. This approach has the advantage of allowing separate team or individual efforts to develop a new system component without jeopardizing the integrity or security of already developed modules. There is a significant cost in the management of multiple processes, but solutions are now offered.

Model Development for Nonprogrammers

Perhaps the majority of software development is concerned with enabling nonprogrammers to interact with computer software and data. Like an artist holding a pallet of colors in front of an empty canvas, a computer programmer with a library of software has unlimited possibilities for creating something new. The artist, in consultation with a client, can create a work that meets the needs or interests of that client. The identical relationship exists between the programmer and a client. The drawback to this relationship is that without the artist or programmer, the client cannot create a new work. An option is for the artist to create pieces of art that the client can arrange at a later time into a final picture. The artist might paint several backgrounds, some pieces of furniture, and some plants, animals, and people in different poses and different sizes. Using this collection of parts, the client can quickly assemble a picture. Of course, the final product will not be as polished and elegant as a work of art commissioned by the client for full completion by the artist, but it is possible for the client to rapidly assemble a finished product—without being an artist. Model development environments have been created for a similar purpose. Programmers develop a general purpose and open-ended set of tools and components that can be applied by a client to create a final product. Examples include word processors, spreadsheets, and graphical software that has a lot of clip art.

A number of example model development environments are available; the number is expected to grow rapidly in the coming years. Currently, a number of different dynamic simulation modeling environments allow us to specify a model, including STELLA[2] and PowerSim.[4] StarLogo[4] and Ecobeaker[5] are examples of spatially explicit simulation modeling environments that are useful for educational purposes. The Spatial Modeling Environment (SME) is a powerful environment that runs models developed with other software (e.g., STELLA) simultaneously on every patch of a raster GIS database.

Considerations to use in evaluating end-user–oriented simulation modeling environments are discussed below.

User Interface

The most visible component of an end-user–oriented software product is the user interface. The challenge is to develop an interface that balances the goals of meeting the needs of a broad array of users, provides a short learning curve, allows access to varying levels of sophistication, and allows rapid access to what a user currently needs.

---

[2] High Performance Systems, Inc.—http://www.hps-inc.com/.
[3] PowerSim—http://www.powersim.com/.
[4] StarLogo—http://StarLogo.www.media.mit.edu/people/StarLogo/.
[5] EcoBeaker—http://www.ecobeaker.com/.

Hierarchical View of a Model

In a business setting, the level of detail that any one person looks at is associated with their position in the organizational hierarchy. Some people pay very close attention to every detail of the development, installation, or support of a product. Many people typically are involved at this level. Managers of such groups must look at how the group will be hired, organized, and supervised. People close to the top of the organization must look out over the future, make strategic decisions, and avoid involvement in product details. Similarly, in model development there should be the ability to visualize the model at varying levels of detail. For example, some depiction of the entire model should visually fit on a single interface screen. A full model might have a number of modules. Those modules might have submodules, and so on. Finally, at the bottom are detailed procedural software commands that the software must easily capture and display.

Unit Tracing

As any algebra student knows, if we do not keep track of the units, we can unwittingly get ourselves into serious trouble. Space probes can get lost if different probe development teams use different units. Tracking of units among different components of a model is a necessity. The author has yet to see a modeling environment that automatically tracks units. Users are left with this unsupported responsibility.

User Constraints on Detail and Time of Processing

Generally, the more complex a model, the better its power of prediction, and the more processing time required. Ideally, a modeling environment should allow for several levels of detail. If a user is happy with a quick ballpark answer, it should not be necessary for that user to run a complex model for a long time.

System Trace-Back

Is it possible, during a simulation run to stop, back up in time, then explore the rules and algorithms that calculate the next step? Sometimes simulation models apparently run fine to a point and then the outputs become questionable. For the sake of understanding and potentially correcting the model, it can be important to be able to trace the model operation at user-specified times in the simulation.

Are Model Components That Seek Equilibrium Points Available?

Examples include carrying capacity numbers, succession steps, logistic growth, and equations of Nicholson–Bailey, Lotka–Volterra, Rozenzweig–MacArthur, and Leslie's predator–prey.

Support Multiple System Levels for Ecological Processes

Does the system support ecological processes occurring at a wide range of spatiotemporal scales? A fundamental notion of hierarchy theory is that different processes happen at different scales in time and space. Simulation of the processes is most efficiently captured if the largest spatial and temporal scales possible are used.

Support Multiple Scales

Ensure that multiple scales in time and space can be simulated simultaneously. Different processes occur at different scales in time and space. At times it is best to reduce a system to simulate smaller components; at other times it is most effective and efficient to clump components and processes. Choice of the proper scale (or scales) depends on the question being asked.

Allow Any Given Simulation Component to Alter Its Operational Time and Space Scales

In certain life stages or at certain times of the day, biotic and abiotic processes are more efficiently modeled and simulated at different scales than at other times. For example, movement of a herd is best simulated at relatively short time intervals over perhaps shorter space during times of migration. Another example: the processes involved in moving water across or through a watershed are relatively slower during dry periods than during wet and changing periods.

Provide for Capture of Processes Within Processes

At times it appears prudent to clump or aggregate ecosystem components; at other times (or to other people) it seems wise to reduce the simulation to its basic components. In the latter situation, it is important to be able to present the reduced simulation as a whole to other system components. For example, it may be important to simulate a herd of animals as a herd for certain portions of a simulation, but as individuals for other portions.

Do Not Force Any Particular Ecological Hierarchy

For example, allow for an ecosystem to exist within an individual as well as an individual within an ecosystem.

Explicitly Recognize the Heterogeneous Distribution of Watershed Components

It must be possible to explicitly recognize and be able to simulate the changes to patterns across a watershed. At any given spatiotemporal scale, some processes and watershed components will be best simulated as a single homogeneous process or entity on the watershed (e.g., weather), while others must be captured as heterogeneously distributed (e.g., slope, elevation, streams, and vegetation stands).

Movement Among Patches

Movement of animals, chemicals, and information among patches must be possible.

Movement of Patches

Patches must be able to come into existence, disappear, and be able to grow and recede in any given direction.

Provide Ability to Analyze Heterogeneous Watersheds

The response of an ecological component to a pattern on the watershed might be accomplished through a brute-force simulation of other entities within the pattern using very short time intervals, or it may be accomplished using a single response to knowledge about the watershed pattern as captured in such measures as provided by percolation theory.

Provide for Simulation of Individual Entities

Although the ability to simulate individual entities is not found in watershed management models, the software advances made by the artificial life community provide a good indication of the latent potential.

Allow System Components to Learn and Evolve

Artificial life systems often enable model components to learn and evolve. In the context of watershed management, evolution is not a significant factor. However, the evolution of behavior patterns based on learning is important where intelligent animals are concerned, but difficult to model in a manner that reflects the natural world.

Use Existing Simulation Models

Significant research and development efforts have resulted in a number of very good simulation models. If at all possible, model integration should first use such models intact.

Support the Movement of Water Across Watersheds

The movement of storm water across a watershed is very important for establishing location-specific erosion and soil moisture content.

Support Simulation of Water Movement Through Stream and River Networks

Once water becomes a part of established streams and rivers, the simulation of the behavior of water in these bodies is crucial to the prediction of flood events.

Accommodate the Subsurface Movement of Water

In many environments, the movement of water through aquifers is critical to the prediction of aboveground plant and animal communities.

Support Scouring, Deposition, and Chemical Transport

At some spatiotemporal scales, it is the movement of large amounts of material via moving water that is critical to the larger system. At others, the movement of chemicals through different soil types, textures, and chemistries is important.

Provide for Simulation of Individual Plants and Trees

JABOWA- and FORET-type models focus on the response of individual plants to the existence and state of surrounding individuals; such models have been well developed and similar capabilities must be available in any general-purpose watershed simulation capability.

Accommodate Simulation of Fire

Fire simulation could be at scales that capture the minute-to-minute behavior of a fire as well as scales that capture the development of fuel buildup over years, the stochastic occurrence of fire, and the effects of fire on the patchwork established on the watershed.

Allow Modeling of Individuals

While the JABOWA and FORET families of forest models provide for simulating forest dynamics at the level of the individual tree, few significant discrete mobile entity simulation capabilities have been identified in the context of ecological simulation. One notable exception is efforts to anticipate military training impacts on a watershed by simulating the anticipated movement of vehicles reported for the Australian Commonwealth Scientific and Industrial Research Organisation (CSIRO) (Cuddy 1993). Facilitating the capture of the behavior of individual animals is the principal focus of the research documented here.

This chapter provided guidelines that can be used to select a model for application in a watershed or watershed management situation. It is most important that modeling requirements and expectations be defined and understood first. This exercise lays the groundwork for selecting a model or modeling environment.

Clearly the evaluation of models and modeling environments can appear costly and difficult. Also, any given evaluation can be out of date within a couple of years as the models and modeling environments are continually improved, updated, and even retired. Different sets of requirements and constraints at any given watershed make it impractical to attempt an evaluation of models in general. This book points the reader to a wide variety of models and this chapter outlined a set of considerations that can be used to evaluate models for local application. The Appendix at the end of the book provides a list of Internet resources for gathering information about models.

Part III
An Integrated Watershed Modeling and Simulation Future

Part II provided a view of the current opportunities for applying dynamic, spatially explicit simulation modeling to watershed and watershed management. Virtually all of the options are specific to a single scientific discipline and, in fact, are derived directly from scientific models. These models were developed to help scientists understand nature; they include overland water flow, stream and river flow, vegetation (natural and agricultural) growth, weather, climate, and plant succession. Nature yields her secrets when science allows only one, or a very few variables to change in response to experimentation, or when enough data are collected to allow statistics to find significant correlations. Typically, each resulting model is able to simulate only some fraction of all of the salient processes at some particular spatial and temporal scale. Unfortunately, the output from several models, each running at different scales and focusing on different aspects of the whole, cannot easily be combined. A current challenge, then, is to integrate several different models in a manner that allows them to run against a single simulation clock and be able to exchange and share information. Linked models that allow interaction among disparate components of the system would dramatically increase modeling realism and overall predictive capabilities.

This part considers watershed management modeling approaches currently under development. In Chapter 2, Figure 2.3 presented a picture of the future in which simulation models developed for scientists have been integrated into watershed management-oriented simulation models. Here in Part III, Chapter 13 reviews a number of different approaches currently underway to realize this future. These appraches range from putting a common interface on a set of otherwise disparate models to creating brand new modeling languages that facilitate tight interconnections between next-generation multidisciplinary model modules. Chapters 14 through 17 conceptualize in some depth the latter approach. Chapter 14 presents a design for a possible future geographic modeling system (GMS), I-STEMS, the Integrated Spatio-Temporal Ecological Modeling System. It is designed to meet the needs of natural resource managers for anticipating the state of a watershed over time based on land and watershed management options, historical and current records, and rules that capture the interaction of watershed and human activities and responses.

I-STEMS will provide the simulation environment within which land management decision support systems will be built for end-user land managers. This section is written in the form of a design document that is focused on I-STEMS' functionality and is designed to meet the future needs of land and watershed managers. The system is approached first from a set of design philosophies based on requirements discussed in previous chapters and then from the viewpoints of basic interface needs of the natural resource manager, the modeler, and the I-STEMS developer.

13
Approaches to Future Model Integration

The preceding chapters have established that there are numerous watershed simulation modeling capabilities developed over the past three decades, but that these efforts tend to be discipline-centric. There are excellent groundwater models, surface drainage models, river water models, transpiration models, plant growth models, and economic models. There are also excellent simulation model development environments that support both the programmer and nonprogrammer. The current challenge is in the simultaneous application of these modeling capabilities to address complex multistakeholder, multidisciplinary watershed management decisions. This chapter reflects on five different approaches currently undertaken at universities and government agencies to make watershed simulation modeling more accessible, practical, cost effective, and immediate to watershed managers. These approaches are

- Common user interface
- Scientific models integrated behind management models
- Scientific models converted to management model modules
- New management models
- New modeling languages

Each approach is discussed separately below. Currently, the most common approach to model integration might be called the professional ad hoc approach (Fig. 13.1). A watershed manager who wants to use simulation modeling to evaluate alternative management strategies contacts a modeler who is typically employed at a state agency or university. This individual selects familiar models and modeling environments to create and run a location-specific set of simulation models. There are advantages and disadvantages to this time-honored approach. First, scientific modelers often argue strongly that a good modeler using a bad model will deliver better results than a layperson using a good model. In many cases this is probably true, but it might be a reflection primarily on the quality of the model's user interface. Perhaps the earliest cars were also best operated by the engineers who created them. Successful operation of the vehicle required knowledge and

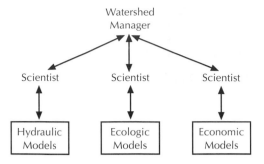

FIGURE 13.1. Common approach to application of models in watershed management

skill in not only the use of the gas pedal, brake, and steering wheel, but also operation of the spark advance, the air–fuel mixture, the shifting of clumsy gears. It also required an enhanced sensitivity of the engine's operation in different temperatures and humidity conditions, the engine's response to different mixtures and grades of fuel, and the vehicle's performance on different road conditions. Perhaps today's state-of-the-art watershed simulation models are not unlike the early automobiles. Made by scientists, for scientists, they ask for and require dozens of inputs. Future versions of these models, like today's versions of yesterday's automobile, will be associated with increasingly complex user interfaces associated with a limited number of user options. Users may have the option to "pop the hood" on these future versions to tweak other variables, but will likely see a sign warning that there are "no user-serviceable parts." The following five subsections review and explore approaches currently underway to create watershed manager–friendly simulation modeling opportunities. The approaches progress from easiest to develop to hardest and from hardest to operate to easiest.

13.1 Common User Interface

In the common user interface approach (Fig. 13.2), existing scientific models are collected behind a common user interface that runs each model individually as intended by the scientific model developers. Output from models is automatically reformatted to allow for easy communication among models. The user is generally not aware that separate software programs are executed behind the common interface. The most common commercial geographic information systems (GISs) through the 1990s were developed with this approach. A common user interface and common database allow a large set of individual processes to operate as an integrated system.

Originally, in GISs, the user interface consisted of a set of commands that could be entered at a command-line prompt on computer terminals. Shell

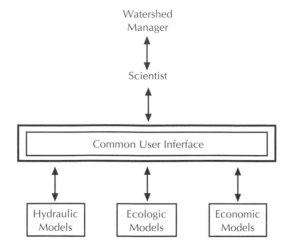

FIGURE 13.2. Models integrated under a common interface

scripts (or batch files) could be constructed to capture a set of operations that, collectively, completed a multistep operational task. As graphical user interfaces (GUIs) developed, these commands were augmented with graphical icons and prompts for user input. In some cases, the original command-line options became unavailable. With this approach, the overall course of operations is conducted step-by-step by the operator. To recapture the ability to put a number of steps into a recallable set of instructions, GUIs were developed that visually linked a number of operations. Figure 13.3 demonstrates this idea. A GIS user assembles input maps and GIS processes (ovals) that together generate temporary maps and a final result. This approach has been used in the Khoros[1] system, the ERDAS[2] image processing system, the GRASSLAND[3] GIS system, and now in the ESRI[4] ArcView 3.2 ModelBuilder application.

A suite of simulation modeling systems developed by the U.S. Army Corps of Engineers has successfully combined the power of proven watershed simulation models behind a common user interface. The Groundwater Modeling System (GMS), Surfacewater Modeling System (SMS), and Watershed Modeling System (WMS)[5] each coordinate a growing number of simulation models behind a common user interface. Users import and prepare GIS and tabular data that is then automatically reformatted so that

---

[1] Khoros—http://www.khoral.com/.
[2] ERDAS—http://www.erdas.com.
[3] GRASSLAND—http://www.globalgeo.com.
[4] ESRI—http://www.esri.com.
[5] GMS, SMS, and WMS—http://ripple.wes.army.mil/software/.

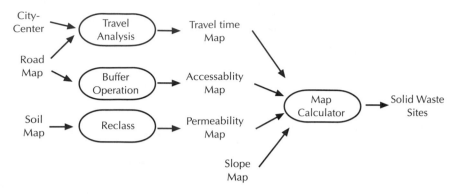

FIGURE 13.3. Graphical environment for process linking

proven simulation models can be executed from the same user interface. Currently, these systems do not employ a GUI as depicted in Figure 13.3, but plans are underway to do so.

13.2 Scientific Models Integrated Behind Management Models

Although the common user interface approach described above increases the accessibility of simulation models through integration behind a common user interface, that approach is still inadequate for being fully attractive for use in watershed management offices. Even with the very nice and consistent user interfaces, the models are still suitable for use only by scientists and engineers. Watershed managers must view the watershed in a fully integrated social, economic, hydrologic, and ecological context. Hence, a decision support system (DSS) that allows the manager to test alternatives with respect to all of these considerations is required. Simulation models are not by themselves suitable as DSSs, but can be used to feed information to such systems. Watershed managers can be provided with DSSs. Such systems might invisibly run simulation models or they can be parameterized by scientists who run simulation models (Fig. 13.4). Parameterization of a relatively simple model through analysis of output from very sophisticated models can be very powerful. Parameterization avoids the significant costs involved with running complex simulation models when a management question is posed.

Multiple, complex, discipline-centric simulation models can be run on the manager's local machine, or can be run remotely across a local or wide area network. A number of research products are currently available and under development to help connect remotely running simulation models.

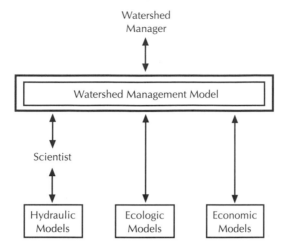

FIGURE 13.4. Scientific models integrated behind management model

WebFlow[6] and Legion[7] allow simulation models to register themselves so that they might be operated through remote requests. FRAMES[8] can provide the GUI on remote machines to allow the integration of various Internet- and local-based models.

This approach, along with the "common user interface" approach, attempts to preserve the legacy of existing, proven software through few, if any, changes to any working software. The primary benefit is that proven, existing simulation models can be tapped as-is, thereby minimizing the cost involved with modifications to the original software and preserving a single lineage in these models. There are serious costs that must be accepted since they can severely limit the acceptance of the final results. First, the original models were not intended to work together. They can use different data definitions and use different partitioning of time and space. Although the original models are typically discipline-centric, they may model components of the system "owned" by other disciplines. A hydrologic model may contain a simple ecological model and an ecological model may contain a simple hydrologic model. Sewing the primary hydrologic and ecological models together can require changes to both models. Second, for systems under continuing development, accepting those systems into an integrated environment can mean a forced adoption of a particular release. As input requirements change (even slightly) and/or output formats are changed, it

---

6 WebFlow—http://osprey7.npac.syr.edu:1998/webflow/.
7 Legion—http://www.cs.virginia.edu/~legion/.
8 FRAMES—http://mepas.pnl.gov:2080:2080/earth/earth.htm.

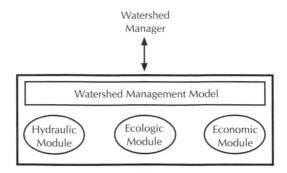

FIGURE 13.5. Scientific models converted to management model modules

will be necessary for the integration team to readapt the new release. This can be a time-consuming maintenance challenge that might not be accomplished. Third, combining simulation models in this manner typically requires that the models be executed serially. This can be inappropriate if the simulation models are significantly changing conditions that other models accept (at startup) as fixed states. To alleviate this problem, models can take turns running so that output from one model can be input into other models on a more timely basis. Alternating control can seriously decrease computational efficiencies as models are repeatedly started and read and write state information to disk.

In the early 2000s, watershed simulation modelers will continue to focus most of their integration efforts in this manner because it allows working software to be maintained with little change while improving the integration of traditional discipline-centric software.

13.3 Scientific Models Converted to Management Model Modules

To address the challenges described at the end of the previous section, a more radical approach might be needed. This approach requires that existing (legacy) simulation models be stripped of their user interface and data inputs and adopt a common execution environment that allows the models to be run synchronously with other models. Models are converted to modules that might be subroutines or separately running processes potentially running on different networked machines (Fig. 13.5).

The Modular Modeling System (MMS),[9] developed by the U.S. Geological Survey, responds to the need for tight couplings in watershed processes

---

[9] MMS—http://wwwbrr.cr.usgs.gov/mms/.

by integrating at the subroutine level. Fundamentally, MMS is a set of matched software subroutines that can be compiled together to represent a particular watershed. The subroutines are called modules and represent rainfall, transpiration, surface water runoff, groundwater, insolation, evaporation, snowmelt, stream flow, and forest growth. To easily assemble the modules to represent a watershed of interest, a GUI allows a nonprogrammer to identify the important processes and how they interact with one another. Weather modules generate rain that is associated with overland water flow modules that empty water into streams that flow into reservoirs and other streams, and so forth. Integration of the modules is managed through a strict control of data definitions and data exchange formats.

The Department of Defense (DoD) has been developing battlefield and theater-of-war simulation models for several decades. The latest attempt to get separately funded related simulations to talk together is High Level Architecture (HLA).[10] Like MMS, HLA requires that simulation modules adhere to common data definitions and data-sharing formats, but it goes further. The High Level Architecture allows the different functions to run separate processes on a computer or on many networked computers. At run-time, cooperating simulation models, called a federation, communicate with one another through a run-time infrastructure (RTI) implementation. The High Level Architecture is an architecture that specifies how federates request and receive information from other federates and how the RTI provides the primary communication channel. How the RTI facilitates this communication is purposefully not specified so that over the coming decades increasingly efficient RTIs might be developed without requiring any rewriting of the federates. Each federation must agree on its own object model, the federation object model (FOM). The HLA effort is ambitious, involves huge military investments in battlefield and theater-of-war simulation modeling, and primarily supports this high-investment enterprise.

High Level Architecture was preceded by the DoD Distributed Interactive Simulation (DIS) environment. Shortcomings of that environment encouraged researchers funded to develop DoD simulation models to improve on DIS. A headquarters-level effort resulted in the development of the HLA specification. One of the researcher-level projects resulted in an alternative called the Dynamic Information Architecture System, (DIAS).[11] Developed at the Department of Energy's Argonne National Laboratory, DIAS has been successfully applied to watershed simulation model development and business modeling. Like MMS, DIAS is associated with a growing number of simulation modules that can be compiled together to match a particular watershed. Like HLA, it allows for the capture of legacy simulation models running as separate processes (poten-

---

[10] HLA—http://hla.dmso.mil/.
[11] DIAS—http://www.dis.anl.gov/DIAS/.

tially across a network) to participate in a simulation model. DIAS has been discussed previously in Chapter 6.

The DIAS approach results in much better management-oriented simulation models, but requires effort on the part of simulation modelers to rework their simulation models. Even after reworking legacy models to work within this more tightly integrated environment, the simulation modules can retain difficulties based on data definition, internal data formats, and related inefficiencies in communication.

13.4 New Management Models

The cleanest approach from a computer science perspective is to develop altogether new software that is integrated up front and avoid the complexity in interfacing software developed at different times and places with different languages and approaches. Geographical information systems developed in this manner. Before integrated GISs, many individual spatial analysis programs were developed using different computer languages, data formats, and programming approaches. Generally useful GISs emerged when large sets of programs were written that used common database formats, common software subroutines to access and manipulate data, and a common and consistent user interface. The challenge is to build a common development environment that is adopted for the development of a large set of integrated software components.

At the foundation of the common environment is a programming language. Ideally, all software in a given system is written with a single language. The number of commonly available languages is a testament to the fact that agreement on a language is not easy. A relaxed strategy involves the development of a core application programmer interface (API) written with a single language, but providing interfaces to a number of different software languages.

A common user interface and database format are required for an integrated watershed simulation modeling system. Another common component is the notion of simulation time. Components of an integrated simulation modeling environment must share a common clock so that the different interoperating watershed systems can remain synchronized through the course of a simulation. SWARM[12] and Spatial Modeling Environment (SME),[13] both discussed previously in Chapter 7, are examples of environments that allow for the development of new simulation modeling code around a common framework. Basically, SWARM is a set of objects written in the Objective-C language that provides a common simulation

---

[12] SWARM—http://www.santafe.edu/projects/swarm/.
[13] SME—http://kabir.cbl.umces.edu/SME3/.

modeling paradigm, user interface, simulation time management, and run-time visualization. Associated object libraries have been developed by a number of different organizations and contributed back to the central SWARM distribution center that further extend a base set of software upon which new simulations can be built. Using SWARM, an Objective-C programmer is free to focus on the development of software that captures the dynamics and state of the system being modeled. Similarly, SME provides a common framework for designing and developing spatially explicit simulation modeling by providing a common simulation clock, a common user interface for controlling and visualizing simulations, and a specific simulation modeling paradigm. The MMS (discussed earlier in this chapter and introduced previously in Chapter 6) also provides a common environment for the design and development of new simulation modeling components.

13.5 New Modeling Languages

A variant of the "new management models" approach is the development of new integrating languages. Academic disciplines have been sufficiently isolated over time that they have developed their own languages, which have been captured in their respective simulation models. Integrating models developed in the milieu of different academic disciplines can have the complication of language (and paradigms) in addition to the purely computer science challenges of languages and approach. Research scientists are beginning to develop standardized modeling languages that allow for the automatic integration of models that were developed within the traditional paradigms of historical disciplines. One attempt at the development of new languages is the Integrated Modeling Architecture (IMA)[14] organized at the University of Maryland. Integrated Modeling Architecture seeks to develop an extensible markup language (XML) that will allow modelers from different disciplines to create discipline-specific models with a language familiar to that discipline. Each discipline will have a discipline-centric development environment, but the resulting models will be automatically linkable through a common XML-based interpretation of the instructions.

Modelers in the early 2000s will be pursuing solutions using all of the approaches outlined in this chapter. Watershed managers should expect that the first efforts will be based on relatively simple approaches to integration of models through automatic reformatting of outputs of some models into inputs for other models (see Fig. 13.4). Reformatting allows currently working discipline-centric models to be run as-is. It also preserves the integrity of working models and is relatively easily accomplished. The

---

[14] IMA—http://swan.cbl.umces.edu/~villa/IMA/.

U.S. Environmental Protection Agency's Better Assessment Integrating Paint and Nonpoint Sources (BASINS) software and the U.S. Army Corps of Engineers' Land Management System (LMS) take this approach. The author expects that this solution will make simulation more accessible for watershed management purposes, but that more intimate simulation modeling will be necessary (see Fig. 13.5) to more intimately link the various system processes. The simpler approach presumes that it will be sufficient to run one model (e.g., the hydrologic model) for a management time horizon and then feed outputs from that model into other models (e.g., crop, habitat, and stream ecosystem). Finally, outputs from these models can be fed into another model (e.g., an economic simulation model). Managers will be forced to conceptually separate the processes (as has the academic community) in this coarse-grained approach. Research efforts over the next decades will evaluate the effectiveness of these coarse-grained models with respect to fine-grained approaches wherein the different components (e.g., hydrologic, ecological, and economic) are run simultaneously, each continually updating a common picture of the state of the system. The next chapter identifies the characteristics of a next-generation management-oriented simulation-modeling software environment. The following three chapters explore such a system from the viewpoint of managers, model developers, and system designers.

14
Design Philosophies

Design and development of a software environment is a complex undertaking from many different standpoints. This chapter introduces design concepts of a hypothetical system that will here be called the Integrated Spatio-Temporal Ecological Modeling System (I-STEMS). In this effort, a large number of design goals must be recognized, considered, and addressed. The goals must be addressed by a collaborative interdisciplinary team that includes target end users, ecologists, economists, statisticians, simulation specialists, mathematicians, and computer programmers. Note that computer programmers are put at the end of this list not to minimize their importance, but to highlight the importance of the other players who can be neglected in software development projects. Key design goals and philosophies are presented separately in the sections below.

14.1 Embrace Current Ecological, Economic, and Management Theories

I-STEMS is about integration; it should not be an exercise in reinvention. This integration is especially important for simulation software in support of ecological, economic, and management components of the system. I-STEMS is designed as a state-of-the-art land management tool. As such, it must recognize and embrace current theories being used in practice today. Although new theories, concepts, and ideas are emerging, the land manager typically is relying on systems based on mature theories that have proven effective. Although I-STEMS can be used to test and develop new theories, its focus will be on the application of concepts and ideas that have proven to have strong predictive capabilities. Drawing from ecological theory discussed previously in Chapter 3, the following development guidelines are suggested:

- Provide for the simulation of individual entities.
- Allow system components to learn and evolve.

- Allow for model components to respond without regard to any predefined equilibrium.
- Allow for the simulation of ecological processes occurring at a wide range of spatiotemporal scales and ensure that multiple scales in time and space can be simulated simultaneously. For example, the raster-based watershed simulation might occur with a spatial resolution of 100 m and a time step of 1 wk while individual animals are simulated at spatiotemporal scales with resolutions of about 1 m and 1 sec.
- Allow any given simulation component to alter its operational time and space scales. The Spatial Modeling Environment (SME) simulation approach fixes the spatiotemporal scale, but the design of the animal-based simulation environment should offer the opportunity to dynamically alter scales.
- Provide for the capture of processes within processes.
- Do not force any particular hierarchy. The general-purpose simulation environment must allow for the development of a wide variety of hierarchies. It must be possible to develop, for example, an ecosystem within an animal as well as an animal within an ecosystem.
- Explicitly recognize the heterogeneous distribution of watershed components.
- Animals must be able to move across the watershed and encounter the differences associated with different locations.
- Standardize the systems integration.

14.2 Use Existing Code

An I-STEMS software development process is very complex and requires a hierarchical approach. While it is arguably true that any piece of software could be improved by rewriting it, reformulating key data structures, and using modern software development paradigms (e.g., "object-oriented programming"), the constraints of time and resources generally will not allow wholesale code rewriting and systems integration. As with any hierarchical system, there will be components that persist because they work very well and reliably even if not especially efficient or elegant. They continue to show up in future generations of the software because the cost of redevelopment outweighs redesign benefits even though individual developers and designers may feel overly constrained because of the continued existence of such components. This argument can hold true for small and very large system components. System integration efforts must allow, for example, for individual systems to be adopted as unchanged whole and complete units. The modification of working code by anyone but the original authors can result in a decreased ability of the original authors to help debug future problems. These problems can include errors in other parts of the system due to a reliance on a particular behavior of the code fragments before modification, and loss of the ability to easily capture future versions of the integrated code without similar upgrades. For further arguments refer to Frysinger et al. (1993).

14.3 Minimize the Number of Authors of Any Given Module

Software tends to be developed as interconnecting parts. The individual parts may be at different stages of development. That is, some parts may be very solidly written, while others have not been fully thought through. Also, software is often asked to accomplish more by its users than intended by its writers, and well-written software should accommodate. In commercial settings that support large user communities, it becomes very important that the thoughts, concepts, expectations, algorithms, and assumptions behind software be thoroughly documented to improve the efficiency of a person who is new to the code and assigned to fix or improve the software. In research settings, however, the lack of a user community makes it possible to experiment more and to actually develop more software in a given amount of time. In these settings, experimentation and development are more important than having solid pieces of software. To compensate for the lack of support documentation, it becomes important that the original authors of software components retain control of those components for as long as possible; that is, if fixes or upgrades are required, the original author (who possesses the understandings of the concepts, algorithms, and assumptions associated with the code to be fixed or upgraded) should be afforded the opportunity to make the necessary changes. If a second author makes changes without the benefit of the original author's knowledge shared through good documentation or close collaboration, the changes may result in a product not fully understood by either the original or the second author. It would not be unusual for a third author, assigned to make changes or improvements, to simply decide to completely rewrite the entire module since rewriting could be judged to be more efficient than taking the time to understand the thinking of the original and secondary authors. Hence, in a research environment, it becomes important for the original author to "retain ownership" of any software developed until that software is fully documented. If full documentation is to occur, it often is not developed until near the end of the associated project.

14.4 Embrace Legacy Software

Existing, proven, and well-appreciated watershed simulation software must be selected for inclusion into an anticipated growing family of I-STEMS software. Embracing legacy software allows an I-STEMS research and development team to focus on a more narrow set of design and development issues. Redeveloping working capabilities does have the advantage of allowing the software programmers to better integrate disparate software, to optimize the algorithms for exploiting a particular set of hardware, and to create better consistency between various pieces of software. These benefits come at not only the cost of redevelopment, but also the loss of potential collaborations with the developers of legacy software. Also, future users will be calling the new development staff with problems that would have otherwise gone to the development team associated with the legacy software that was not used. Finally, legacy software typically has developed a following of individuals who form a pool of new customers who have been satisfied with the results from that software.

14.5 Design Everything to Be Modular

Today, software design and development embraces modularity. Object-oriented programming defines today's preferred approach to modular programming. For I-STEMS, modularity is a goal for all levels of design. The system will rely on only a small set of components. Most will be optional, replaceable, and interchangeable with other components. How they are integrated will depend on the particular simulation being developed for a particular end user.

Strong standards will be developed to ensure the mixing of different components developed by different I-STEMS teams. If a particular graphical user interface (GUI) is to be effective as a viewer or controller for any number of subsystems, those subsystems and the GUI must be developed to a common set of specifications. A cornerstone of these specifications will be the requirement for system components to be individually functioning objects that run as separate programs on a network yet connected to other objects through interchanges of information across the network via standardized protocols.

14.6 Allow Distributed Processing

Dynamic, spatial, ecological simulation models can rapidly become very complex and can overwhelm single-processor computer systems. It is important that the simulation models assembled to address real watershed management questions and concerns be able to use any number of available processors. These processors may exist within a single machine or may be distributed across a heterogeneous network. Distributed processing can be accomplished through assorted means. First, as noted above, I-STEMS system components will effectively be independently operating programs communicating with one another in a heterogeneous network of comput-

TABLE 14.1. Three levels of system interface.

Interface Level	Activity	System View
Software developer	Encapsulation of legacy software into subsystem objects Development of new simulation software modules	Application programmer interfaces Software modules that run as stand-alone programs
Model developer	Develop models for end-user resource managers	Libraries of system components Configuration files Assorted viewers and controllers
Resource manager	Manage watersheds with respect to mission goals	Any number of simulation models A consistent interface among models

ers. Because the components operate as single programs, they can be easily distributed across any number of processors and computers. Second, some modules will be developed that make use of specific parallel processing environments as well.

14.7 Allow Multiple Interface Levels

I-STEMS will be developed with at least three user interface levels in mind (Table 14.1). The I-STEMS software developer will work with well-defined application programmer interfaces (APIs). These interfaces will include all of the standardized system objects, routines, and data exchange methods to support the encapsulation of legacy software simulation components as well as the efficient design and development of new components. Model developers will then use the I-STEMS modular model components to create location- and management-specific models to be used as watershed decision support systems. The models they create will be used by land managers for risk assessment, analysis of impacts, and improvement of land management techniques and schedules.

14.8 Design Model Components as Objects

I-STEMS will embrace object-oriented software design approaches. Design and development of objects is more expensive than traditional programming approaches. Also, execution time for software developed with objects can be significantly slower. The payback occurs with the ability to rapidly recombine sets of objects. Because each object is essentially a self-contained program, it can be combined with other objects without worry about conflicts with other software. Each appears to the other objects as a "black box" with required inputs and available outputs. The internal operations by the object are hidden from other objects.

To the future model builder, I-STEMS will offer a family of independently developed simulation objects, including any number of watershed simulation modules that may be object encapsulations of pre–I-STEMS simulation models. For example, geographic information system (GIS) operations, hydrologic simulations, plant succession models, and weather simulations will be captured in such modules. I-STEMS will be an open software environment within which any research group will be free to design and develop additional simulation modules.

Encapsulation of models and simulations will be accomplished in a very rigorously defined manner that will ensure that every module will be as broadly recognizable to other objects as possible. The consistency in external appearance of simulation modules will then allow the design and development of user-oriented visualization and control objects.

This chapter provided an overview of major design philosophies involved with the design and development of next-generation watershed-oriented simulation modeling environments. A serious software development group will further refine and develop the ideas presented here to eventually create a design architecture that provides a comfortable balance between manager needs and programmer constraints. The next chapter begins to view the final system from the standpoint of the watershed manager.

15
Watershed Manager's View

15.1 System Design Philosophy

The view of the Integrated Spatio-Temporal Ecological Modeling System (I-STEMS) presented here describes what land managers will encounter when they use the model. I-STEMS will appear to the land manager to be an array of land management decision support systems (DSSs) seen through a familiar I-STEMS interface. These systems will be known as I-STEMS models. Some models will be relatively complete simulations that simultaneously capture all aspects of the watershed processes. Other models will focus on very specific management questions. A large model might simultaneously simulate watershed activities running at a number of different spatiotemporal scales to simultaneously capture such components as the behavior of individuals representing a threatened or endangered species; the behavior of larger populations; human activities, including training, logging, and recreation; economic consequences; biodiversity consequences; fire; disease propagation; and movement of genetics.

Models will have a similar look and feel because they will be constructed from a common toolbox of software. Models will be controlled and viewed through a set of standardized human–computer interface components. Once a manager has become comfortable with a particular model or two, other models will automatically feel familiar.

Human activities occur at a number of different scales in time and space and they interact with nature at each scale. At one end of the land management spectrum is fire-fighting, which involves responding to system breakdowns and shifts in nature that occur at relatively short time scales (hours to weeks). At the other end of the scale, the land manager must consider how human activities and schedules interact with long-term natural activities and how the human activities and schedules affect biodiversity, conservation, and global warming. I-STEMS provides tools and functionality that allow for the construction of models at both extremes.

15.2 Multiple Models

It is anticipated that land managers will eventually use a number of different models to help them manage watersheds. The ideal situation, of course, would for a land manager to have a single model against which any land management scenario could be evaluated with respect to all important consequences. Let's do some "blue sky" musing for a moment and describe this ideal system. First, we need to identify the types of land management decisions that the system will need to evaluate. For example:

- Layout of buildings, roads, and other land uses
- Schedule of land use
- Schedule for land rehabilitation

Next, we identify the kinds of questions that a land manager will want to pose to the system. The following list may cover the range of questions:

- Project the land cover anticipated during a season . . . during a decade.
- Project and evaluate the biodiversity anticipated over a century.
- Anticipate the cost of each activity within the watershed with respect to environmental damage.
- What is the anticipated impact on threatened and endangered species (TES)?
- What is the burn potential for the land during the year?
- Anticipate the erosion potential associated with each activity in the watershed.
- Analyze the watershed's resistance and resilience to disturbance, including fire, disease, storms, and flooding.

These decisions require analysis at a number of different spatiotemporal scales. It is anticipated that a small number of models might be developed to cover the range of questions and objectives. Each model would focus on processes that occur at similar time and space scales. For example, I-STEMS might be used to develop the following modeling systems.

Emergency Simulation and Analysis (ESA)

This system would focus on rapidly changing dynamics associated with emergency situations. It might have the following subsystems:

- Wildfire simulation
- Chemical spill simulation
- Storm and flood simulation

Scales:

- Time step: minutes
- Time extent: days
- Spatial resolution: 1 to 10 m
- Spatial extent: subtraining range to local

These models would be run by environmental office personnel to help guide emergency responses to unusual situations. ESA would initialize a simulation by extracting, from ISSIS (see below), the current state of the watershed. ESA would also be used to simulate the potential for an emergency situation.

Installation Seasonal Simulation and Information System (ISSIS)

A military installation management environment, ISSIS could provide a simulation environment that would be used daily to (1) keep track of the current state of the watershed, and (2) project the state of the watershed over the current season.

Scales:

- Time step: 15 minutes to 1 day
- Time extent: 1 to several years
- Spatial resolution: 10 to 100 m
- Spatial extent: local

Inputs:

- Range control
  Training schedules
  Tables of organization and equipment (TOEs)
- Environmental office
  Land rehabilitation
  Impact model input
  Measurements of watershed health

Outputs:

- Land cover predictions
- Environmental cost of each training exercise
- Anticipated erosion potential
- Comparison of different potential schedules
- Anticipated impacts on TES and critical habitats
- Changes to habitat suitability indices for selected species

This system would be developed for use by personnel in the environmental and range control offices. Each office would be responsible for managing certain inputs. Any office could then use ISSIS to project the watershed into the immediate (1-yr) future. ISSIS would need to interface with other management systems in daily use such as the local geographic information system (GIS), the Range Facility Management Support System (RFMSS), and others.

Integrated Regional Effects Simulation System (IRESS)

This hypothetical simulation model focuses on the long-term (years to centuries) consequences of land management patterns. It would be used primarily by environmental offices to address long-term consequences of land-use patterns, forestry, and regional watershed patterns with respect to biodiversity, sensitive habitats, TES, and successional states of the land.
Scales:

- Time step: 1 month to 1 year
- Time extent: decade to century
- Spatial resolution: 100 to 1000m
- Spatial extent: local to regional

Inputs:

- Land use
  Land use patterns
  Anticipated activities (in time and space)
  Forest management plans
  Ecosystem response models
  Successional models
  Land condition trend data

Outputs:

- Watershed successional state projections
- Long-term TES and Habitat Suitability Index (HSI) potentials
- Biodiversity predictions (regionally oriented)
- Comparison of different potential schedules
- Anticipated impacts on TES and critical habitats

Each of these hypothetical systems would be constructed within I-STEMS, but the users of the systems will not actually be working with I-STEMS. Because the systems are developed within the same environment, they will provide consistent interfaces to the end user.

When standard models (e.g., ESA, ISSIS, and IRESS above) are not sufficient, land managers can turn directly to I-STEMS to design and develop a new simulation model.

15.3 Model Modification

Simulation models run by land management personnel will be associated with a large number of input options. These can be divided into initialization and run-time parameters.

Initialization Parameters

Watershed simulation models must be initialized with the starting state of the system. The state data will likely involve:

- Landscape maps that identify such things as vegetation type and density, topological information, and land ownership
- Schedules of watershed activities
- Weather statistics
- TOEs
- Training activity descriptions

Run-Time Parameters

When models are run, a number of options can affect the run. These include:

- Assignment of subprocesses to computers
- Identification of how the model will be visualized during and after a simulation
- Identification of run-time input options
- Debugging output options

The main message of this chapter is that I-STEMS is conceived to be a modeling environment that allows for the creation of location- and management-specific simulation models that can be integrated into the decision-making process of specific watershed management offices. Any number of different systems should be able to be constructed so that decisions can be quick and efficient through the reflection of local needs and situations in the user interface. I-STEMS itself is the generic environment within which specific models and DSSs can be constructed.

16
Modeler's View

16.1 Audience

This section describes what the individuals developing new simulation models will see when working with the Integrated Spatio-Temporal Ecological Modeling System (I-STEMS). The development of landscape simulation models requires the coordination of an interdisciplinary group of individuals. It is presumed that these individuals will not, for the most part, have the skills necessary to design and develop new simulation modules using low-level software languages. They will, however, be assembling simulation modules into complete watershed simulation models. Through a story, the next section, Imagine, describes the development of a simulation model by an interdisciplinary team. This is followed by sections on the system design philosophy as viewed by a modeler, subsystem examples, and generic viewers and controllers.

16.2 Imagine

To help visualize the utility of the I-STEMS geographic modeling system (GMS), this section develops a hypothetical future scenario that involves a simulation challenge at a military installation. Fort Hood, Texas has been challenged to expand their training areas into adjacent properties. This expansion is desired to accommodate an anticipated expanded training mission for tracked vehicles. The environmental office is tasked with generating several annual training scenarios and then evaluating each with respect to the direct and indirect impacts on (1) the ability to train, (2) the effects on local golden-cheeked warbler populations, (3) the effects on local black-capped vireo populations, and (4) the effects on local and regional biodiversity. This effort is part of the environmental assessment (EA) requirements. Management decides that the analysis shall be accomplished by an interdisciplinary group consisting of individuals from the environmental, training, and scheduling offices. They will have at their disposal several workstations that have recently been used to test the latest version of I-STEMS.

Day 1—Team Assembles

A high-priority meeting is held to assemble and brief the team that will be handling this assignment. They are tasked to develop a dynamic training area simulation model focused on the intended expansion area and adjacent existing Fort Hood properties. The resulting model will be used to evaluate the direct and indirect environmental impacts expected to be associated with the new training area. Known concerns that will need to be addressed before the new area can be used for the intended training include: (1) two threatened or endangered species, (2) potentially sensitive ecosystems and habitats, (3) water quality requirements for drinking water wells downstream, (4) effect on regional biodiversity initiatives, and (5) timber harvest goals. The team must provide a working simulation model and preliminary results within 20 days; the model and results will be presented to visiting dignitaries at that time. In addition to the workstations at each member's desk, the main server located in the environmental office is available. It is a $50 K machine containing 1 Gbyte of internal RAM, has four 200-GHz processors, and 20 Gbytes of on-line hard disk. Fort Hood has been connected to the Internet since the mid-1990s and now has a 100-Mbit/sec connection to the Internet, which provides them with powerful run-time access to several supercomputer centers, including the thriving National Center for Supercomputing Applications (NCSA) at the University of Illinois at Urbana-Champaign. The team will be using the latest release of I-STEMS.

Day 2—Establish Subteams

Following the initial briefing, the team meets and establishes the following subteams:

- Species-specific models
- Weather and climate
- Hydrology
- Communities and ecosystems
- Geographic information system (GIS) and image processing
- Visualization and control
- Training

Each team is tasked with identifying and evaluating available model components. Each team spends the day discovering model components distributed across the Internet and from local sources.

Day 3—Available Component Reports

The simulation team meets to brief each other on the information discovered during a day of exploration. Potential system components are presented in Table 16.1. All components conform to the I-STEMS standards, which allows them to be readily integrated. Team reports are as follows:

- Species-specific team: Three models of local threatened and/or endangered species are available. Population- and individual-based models are available for the black-capped vireo and the golden-cheeked warbler. The team recommends adopting the population-based model.
- Weather and climate team: Weather and climate models have both been identified on the Internet. Both are identified as standard, accepted models and model outputs.
- Hydrology team: The Saghafian (Saghafian 1993) model was located in I-STEMS format. It has now been verified on a wide variety of watersheds. Also, a new soil compaction model conforming to I-STEMS has been located at a server at the Engineer Research and Development Center's Geotechnical Laboratory.
- Communities and ecosystems team: The standard Army Corps–developed plant succession model developed by the U.S. Army Corps of Engineers is available in Beta release Version 4.2.
- GIS and image processing team: Extensive historical and current GIS and imagery data exist on-site and can be adapted to I-STEMS.
- Visualization and control team: The Internet server at the University of Illinois currently offers a wide variety of visualization and control objects for I-STEMS applications. These include the traditional meters, sliders, menus, feedback panels, dials, and buttons. Several sophisticated new intelligent controllers are now also available at this site; they manage trade-off options, various optimization approaches, and collaborative modeling tools.
- Training team: Two training model sets are available. Fort Hood's training impact tables have been used quite successfully for the past decade. They relate training exercises and training areas with degree of estimated environmental damage. The Construction Engineering Research Laboratory's (CERL's) relatively new set of maps adds a spatial dimension to these tables and provides impact information at a resolution of 30 m. The team decides to adopt these maps and the CERL approach to developing such maps.

Day 4—Register Available Submodels

A new model is established on the server. This process consists of setting up an information exchange server that will facilitate communications among different processes running on different machines. All participants are told to set up I-STEMS environments on their individual workstations that attach to this server. Once this is accomplished, any team member can

TABLE 16.1. Hypothetically available model components.

Potential Component	Source	Description	dT,dS	Inputs Required	Outputs Available
Black-capped vireo	CERL	Population model object developed for Ft. Hood (1998)	1 wk, 1 km	Weather Topology Vegetation (grass, forb, shrub, tree)	Densities in 6 age classes
Black-capped vireo	U of Texas	Individual-based object developed for the State of Texas (2001)	1 day, 100 m	Density of predators Weather Topology Vegetation (5 species)	Location Health indices (5) Age, sex, etc.
Golden-cheeked warbler	Texas A&M	Population model object developed for Ft. Hood (1998)	1 wk, 1 km	Weather Topology Vegetation (grass, forb, shrub, tree)	Densities in 6 age classes
Vegetation density maps	CERL	Vegetation, grass, shrub, forb, and tree (2002)	N/A, 30 m	N/A	N/A
Vegetation succession model	Colorado State Univ.	20-species succession model (1999)	1 mo, 100 m	Soil type Soil compaction State of starting vegetation	Succession phase
Tracked-vehicle impact model	WES	Soil compression model (1997)	N/A, N/A	Tracked-vehicle days per ha Soil type	Soil compression
Biodiversity model	INHS	10-keystone species model (1998)	1 yr, 10 km	Climate % land in each of 5 succession states	Densities and genetic variability for each species
Training models	CERL/Ft. Hood	Maps created for each exercise and training area combination (2003)	1 day, 30 m	Training exercise Training area	Average tracked-vehicle days per ha
GIS	Ft. Hood	Extensive 100+ theme digital map data base	N/A, 5 to 100 m	N/A	100+ themes, some historical data; extensive imagery
Hydrology	CERL	The Saghafian finite-difference model (Saghafian 1993)	min to days, 30 m	Topographic data, land use and cover	Saturation, depth, velocity, scouring, and deposition
Weather	National Weather Service	Historical and average weather conditions and probabilities	1 day, 100 m	Day of year	Temperature and rainfall: average, standard dev., and probability

readily query and view any portion of the developing model as well as establish model components on their own machine. Team members then begin to set up the submodels selected from Table 16.1 on the local machines. By the end of the day, each member is able to view the status of the virtual interconnections among the various submodels. For example, a query on the status of the hydrologic simulation model reveals the following report:

Hydrologic Simulation

Main Model Information

Name:	Ft. Hood extension simulation
Main Server:	env.fthood.army.mil
Access Code:	175

Submodel Information

Name:	Hydrologic simulation
Model Server:	hydro.fthood.army.mil
Access Code:	180

Metadata

Author:	Bahram Saghafian
Version:	4.3.1
I-STEMS version:	2.6
Resolution:	30 m

Inputs

Name	Units	Initiated By	Supplied By	Converter
Elevation	m	GIS	N/A	N/A
Slope	degrees	GIS	N/A	N/A
Initial saturation	mm	GIS	N/A	N/A
Soil permeability	mm/day	GIS	N/A	N/A
Manning's K	K	GIS	Vegetation model	N/A
Water	mm/hr	N/A	Dummy	mm/inch

Outputs

Name	Units	Used by Submodel
Soil saturation	mm	Vegetation
Water depth	mm	Vegetation
		Golden-cheeked warbler
		Black-capped vireo
		Training
Water velocity		Vegetation
Soil scour/deposition	mm	Vegetation, succession

This report begins by indicating that the hydrologic simulation submodel has been registered with a "main model" called "Ft. Hood Extension Simulation," which is registered on the machine called env.fthood.army.mil. Connection to this model is accomplished with access code: 175 (which is a port or socket type number). This submodel has registered itself with the main model and will be running on and accessible through hydro.fthood.army.mil. Note that all submodels may run on separate machines. Underlying information brokers facilitate virtually seamless integration of these submodels. Some model metadata are also displayed. Here, that information identifies the version number of the submodel and the latest I-STEMS version under which the model is known to operate. Sections on inputs and outputs provide information on how the submodel is currently linked to other submodels. These links were established using user interfaces that probe the model space for available variables and then allow the modelers to establish the desired connections. The CONVERTER column in the inputs section identifies which, if any, standard unit converters were used to establish the connection. In the outputs section, the submodels that currently use the available outputs are identified. The lists of such submodels will grow and shrink as the different components link themselves up with each other.

The input "Water" is identified as being supplied by submodel "Dummy." This is a reserved submodel name that is attached to very simple data generators. Model developers are allowed to create dummy inputs defined by fixed values or graphs that use time (e.g., month) as the independent variable. The purpose is twofold. First, inputs that are not being generated by other submodels can be simply accommodated in this fashion. Second, during debugging and sensitivity analyses, input variables can be set to static values.

Each submodel can be probed in a manner that results in this type of report. Components other than submodels can be established and then viewed through this report. The most important object classes are viewers and controllers. Viewers are essentially submodels that access output only from other models; they probe submodels and display information many different ways. Viewers can provide run-time views of system states (maps, tables, strip charts, etc.) or can dump data to output files for later analysis. Similarly, controllers are basically submodels that provide input to other submodels. Based on human interactions with graphical user interfaces (GUIs), controllers supply values to submodels. Such inputs are injected into the associated submodels (typically the receiving submodel controls the data probe).

Of the numerous other interfaces available to the modelers, two require a brief mention here. A main control panel is available for starting and controlling the model as a whole. This interface allows the user to change any of the various submodels into ON, OFF, and STATIC modes. OFF makes the submodel appear to be nonexistent. STATIC turns the submodel off, but allows it to generate predefined static information much like the "Dummy" submodel. ON causes the submodel to operate normally during the course of a simulation run. These modes are likewise used to control the view and controller components. The second general type of important interface is the control panel for supporting simple modifications to each of the submodels and view and controller interfaces. For example, a generic population submodel can cover a wide range of populations by simply allowing the modeler to "tweak" such attributes as growth rate, consumption rate, fecundity rate, and home range size; or a user interface might allow a wide variety of displays for a given series of data: bar chart, strip chart, colors, or ranges.

Days 5 Through 10—Research to Develop Missing Components and to Extend or Modify Available Components

The team uses a full week to follow up the initial assembly of available components with development of additional simulation components. In particular, the available visualization tools had to be assembled in a manner that maximized the match to the current application. The training submodels needed to be upgraded a bit to reflect the new training scenarios and weapon systems anticipated for the new watershed. Each submodel is run independently to identify as many potential errors as possible.

It is also decided that two models would be developed to help address the overall goals and objectives. The biodiversity questions require a time step and resolution sufficiently different from the other questions to warrant a separate model. Outputs from the two will modify each other as the same engines are driving the underlying systems.

Days 11 Through 14—Integrate and Debug

This week's effort involves numerous runs of the simulation model with more and more components turned on. As conditions are discovered to move out of reasonable ranges (negative populations, temperatures over 150°F, and succession stages out of line with simulated training), errors in the submodels and data are discovered and repaired. Sensitivity analyses are conducted on the more uncertain inputs—some of which are found to be quite important. The developed user interfaces are also tested and improved to remain stable.

Days 15 Through 20—Management Evaluation of Alternatives—Reports Generated

During the final phase, management representatives are invited to participate in the final simulation runs. Some different training schedules are run along with some updates to potential property boundaries and road network possibilities. Output videos are generated for playback in future meetings and are captured for viewing on the Internet. It appears that more of the objectives than first imagined can be met through newly recognized arrangements of the planned training activities. Key locations, thresholds, and leading indicators have been identified for particular monitoring as a strategy is implemented. The models are documented and made available to the management team for use in making decisions within the chosen strategy.

This imaginary scanario, based in some fact, suggests that a GMS will be useful to landscape managers (here military installation training range managers) for the rapid design and development of location-specific dynamic simulation models. These models will simulate various components of the landscape simultaneously using appropriate spatiotemporal scales for each. Long-term and indirect effects and interactions among the various components will be able to be explored by the managers.

16.3 System Design Philosophy

The previous story suggests a number of design philosophies that are explicitly stated here. First, the modeling environment supports difficult *collaborative* efforts. Specialists are each assigned to peruse libraries of models compliant with I-STEMS to analyze and identify potentially suitable submodels. These submodels will have been designed and developed by specialists (e.g., hydrologists) for the purpose of being later connected with submodels developed by other specialists (e.g., range or plant succession scientists). Submodels reflect the philosophy of *modularity*. Each submodel used in the story was developed outside of the final model being assembled. Each is a stand-alone object that is prepared to behave and interact with other submodels developed at any number of research and development sites.

From a computer science standpoint, the assembled model runs in a *distributed* and perhaps heterogeneous computing environment. Individual submodels will be allowed to run on platforms for which they were developed while simultaneously interacting with other submodels running on perhaps different central processing units (CPUs) and even different machines within a local or wide area network. Each submodel will be developed to communicate with other submodels using *standardized intercommunication protocols*. One class of submodels will be *viewers* and *controllers*. Submodels query the simulation space outside themselves, process that information according to internal rules, and then potentially communicate internal information back to the outside. A viewer is an instance of a submodel that queries the outside and then "internally" projects the results of the query to a human operator through run-time visualization or by saving the data in a file for later viewing and analysis. Controllers, like viewers, communicate with people. Human operators input information into a controller that in turn uses the information to effect some state change in another submodel.

16.4 Model Control Center

An I-STEMS model consists of a number of key components, each potentially running on a different CPU or a different machine on the network. From the modeler's viewpoint these components can be grouped into the following two categories:

- Model control center
- Model subsystems

The control center is discussed here while the model subsystems are described, from the modeler's viewpoint, in the next section. A control center consists of a number of interrelated programs that together provide the environment for initializing and managing the subsystems. It is associated with a user interface that provides various viewports into the operation of the full model. The control center has two primary responsibilities. First, it maintains information about the various submodels being used. This includes model name, machine to which it is assigned, data it requires for initialization and input, and data it can provide during a simulation. For example, Tables 16.2 and 16.3 display sample input and output information for two hypothetical submodels. If these two submodels were instantiated by the control center, Table 16.4 Submodel Data Exchange, could be automatically displayed to identify to the modelers the current match-up between required inputs and available outputs. Associated with each data stream will be data units, associated error information, and frequency of data change. Second, the control center will monitor system and model performances during a simulation. Monitoring may include CPU usage statistics on the various machines and rate of data exchange among submodels (especially among machines across a network).

TABLE 16.2. Hydrology submodel (sample).

	Submodel Hydrology Name:	
	Variables available for output	
	Name	Units
var	Water depth	cm
var	Soil saturation	%
	Variables required for input	
type	Name	Units
fixed	Soil permeability	
fixed	Soil depth	m
var	Manning's K	K
var	Rainfall	mm

TABLE 16.3. Vegetation submodel (sample).

	Submodel Vegetation Name:	
	Variables available for output	
	Name	Units
var	Percent live veg cover	%
var	Percent dead veg cover	%
	Variables required for input	
	Name	Units
var	soil saturation	%
var	High daily temperature	°C
var	Low daily temperature	°C

TABLE 16.4. Submodel data exchange.

Model variable	Initialized by	Managed by	Used by
Water depth	?	Hydrology	
Soil saturation	?	Hydrology	Vegetation
Soil permeability	?		Hydrology
Soil depth	?		Hydrology
Manning's K	?		Hydrology
Rainfall		?	Hydrology
Percent live veg cover	?	Vegetation	
Percent Dead veg cover	?	Vegetation	
High daily temperature			Vegetation
Low daily temperature			Vegetation

Control centers will allow modelers to assemble model components while monitoring the interactions possible among submodels. This will be accomplished in networked environments by "slaving" remote control centers to a master control on a selected computer. Each control center will have access to local tables of available submodels and will be able to instantiate these models as directed by the user operating the master control center. As different submodels are "brought up," they announce their data requirements and offerings, which the master control center manages and optionally provides to the operator. Once a set of submodels is initialized and has all of the input data requirements accommodated, the control center can set and then start and stop the master simulation clock. During simulation runs, optional viewports may display run-time statistics.

Finally, control centers will have the option of saving the parameters associated with a simulation (complete or incomplete) in master files. These files can be used to fully initialize a simulation model at a later time with minimal user interaction.

16.5 Subsystems

As noted above, a fully operational I-STEMS will provide a toolbox of submodels designed and developed at numerous research and development centers, and accessible through libraries constructed on the Internet. Each submodel will be associated with metadata describing not only the characteristics of the model, but also the refereed reviews of the model, identifying conditions under which it is and is not useful. Next to a growing library of submodels, the I-STEMS "core" will be relatively small—providing only standards for submodel design and development that will ensure interactions with other models through well-defined communication channels.

Common "Appearance"

Each I-STEMS–compliant submodel will interact with other submodels via standardized protocols captured within an application programmer interface. To the model developer working directly with compliant submodels, this means that each submodel will offer a common and consistent appearance to other model components. For example, when a submodule is initialized at run-time, it registers (with the associated control center), information that it requires to operate as well as the information it can provide. Information may then be readily displayed by the control center covering all participating submodels (e.g., see Table 16.2).

One set of model components will be a standard set of viewers and controllers. These are discussed in some detail later in this chapter. Because of (1) their repeated use between models and (2) their being the only interface between people and the submodels, the viewers and controllers will provide the most visible consequence of consistency in "appearance" of submodels to each other.

Submodels Are Software Plus Data

Submodels will be designed and developed as independent objects. An object here is defined as a stand-alone set of data and software instructions. Many software developers have adopted object-oriented software approaches, resulting in the development of a wide range of software programming languages such as C++. I-STEMS brings this development paradigm to the model assembly level.

For those readers unfamiliar with object-oriented programming, let's explore that meaning and its consequences. Many I-STEMS model developers will be familiar with the use of GISs. These systems have traditionally distinguished between the data associated with a particular watershed and the software (the GIS) used to create, display, and manipulate that data. To query or analyze a map (or maps), the GIS operator would invoke the local GIS to perform the query or analysis. An object-oriented GIS would combine the operations with the maps. This combination would "exist" on its own—separated from other processes. A person could then ask the object to perform the desired query or operation on itself.

For example, using traditional GIS reasoning, a user would ask a GIS package to invoke a particular program with certain user inputs on a map (or set of maps). The user might start the GRASS GIS and run a command such as.

R.Info Soils

This is a request to run the r.info program on the map called "soils." In an object-oriented GIS, the syntax is reversed. Instead of asking the r.info program to process the soils map, the soils map is asked to provide information. The command might be:

ASK Soils TO GiveInfo

The key difference is that what had been an inert piece of data is now an active entity, capable of responding to certain requests. This changes the way data are viewed and opens the opportunity to integrate digital watershed information in more dynamic (rapidly changing) ways. In a traditional GIS, a process is invoked on a map. This requires that, at a minimum, the map be read into memory, the data be processed, and the map be written back out to disk. Each procedure performed on the map, regardless of the complexity, goes through these steps. If there are to be many operations on a map interspersed with operations on other maps or files, the concept of a map as a "living" object becomes attractive. A map object can keep a map active for the duration of a simulation or for a series of different operations. A simple map object may be able to pull a map into virtual memory and then respond to requests for changes to or information about the map over time.

This object-oriented concept is especially useful for dynamic watershed simulations. Essentially, each I-STEMS submodel is associated with the state and management of certain watershed information—certain maps. Conceptually then, I-STEMS is a dynamic simulation-focused, object-oriented GIS. The model developer will think about watershed simulation as the assembly of interacting dynamic maps. For example, the vegetation-cover-map is actually a dynamic simulation of the vegetation and might use rules based on Clementsian plant succession. A training-simulation-map might represent a training exercise complete with its mission, materiel complement, and limitations on fuel, time, and allotted environmental impact. The I-STEMS model developer must think of submodels as objects that combine behavior rules with system state information. The processes and the data are combined into objects that will grow into extensive libraries.

16.6 Viewers and Controllers

One class of I-STEMS objects will become very familiar to any I-STEMS modeler and, effectively, any manager running complete I-STEMS models. This class contains the viewer and controller objects. As discussed above, an object in object-oriented programs consists of information (data) and operations. Objects respond to requests from external objects and may invoke requests on external objects. In the case of viewers and controllers, human operators provide the "operations"; that is, conceptually, the human operator is viewed by other system objects as existing inside the viewer and controller object. Actually, objects have no knowledge of what goes on inside any other objects; they only knows that they can make certain requests of other objects. That a human or computer automaton resides in the object is of no consequence to other objects.

Viewer and controller objects in I-STEMS provide the only user inter-face. The I-STEMS rule will be that software within model objects does not drive peripherals (monitors and keyboards). The reasons for this are three-fold. First, consistent user interfaces can be better maintained and managed if each submodel is not allowed to perform its own interface with opera-tors. Second, of all software written, the user interface has the shortest relative life span. I-STEMS submodels are expected to enjoy 10- to 30-year life spans while the interface software is expected to be viable for only 5 to 10 years. By forcing the separation of the user interface from the models, I-STEMS will be more efficiently upgraded over time. Finally, software must not be written to require specific peripherals. By forcing submodel devel-opers to use established viewers and controllers, the developers are less likely to write system-specific software.

Four classes of viewers and controllers are suggested below. There are, however, potential combinations of these and others as well.

Run-Time Visualization

A pure visualization object submodel simply probes, during a simulation run, certain user-selected information available from the submodel objects. The methods (software calls) that visualization submodels use are identical to the calls submodels use to query one another. Run-time visualization objects will be developed to provide a number of viewports into the operating model. These will include:

- Map views that might overlay raster, vector, and point information
- Time-series views of selected state variables in "strip chart" formats
- Tabular views of selected state variables
- Views of overall system status, including the load on computational circuits (CPU, network, memory, and disk)

Run-Time Control

Control submodels provide operator run-time input options to a simulation. Simulation control will include:

- Control of the overall simulation. This might include the ability to start and stop portions of the simulation or the ability to exchange one submodel for another.
- Control of individual submodels, such as adding or deleting components, or changing the state of the submodel.

Perhaps most control submodels will also be viewer submodels, for viewing and controlling are conceptually two sides of the communication process. It will not be unusual for a viewer to be used without an attached controller, however.

Data Storage

As a simulation runs, the state of the simulation is continually changing. A complex simulation typically cannot retain the complete state of its system throughout the entire simulation run. Consider a watershed represented by a 1000×1000 grid of cells. Each cell manages 20 state variables and the simulation runs for 500 years at 1-week time steps. Assuming all variables are represented with 8-byte floating-point values, the entire simulation could generate 4.16 Tbytes of output (1000*1000*20*500*52*8). A data storage submodel would act like a visualization model that probes the simulation for the state of the system. However, instead of graphically rendering the output, it stores selected portions of the simulation in files for later statistical analysis.

Postanalysis

The data storage submodels capture data for later analysis. I-STEMS will rely on available data analysis and display software for these analyses. Software will include statistical packages, geographical information and image processing systems, and standard graphics depiction tools, including translators and movie viewers.

This chapter outlined the I-STEMS view from the standpoint of the modeler. I-STEMS is conceptualized as a generic environment containing a broad set of tools for developing and testing simulation models. Model developers will be able to create location- and application-specific multi-disciplinary simulation models by choosing, combining, and parameterizing specific model and user interface modules.

17
Programmer's View

17.1 Audience

The view of the Integrated Spatio-Temporal Ecological Modeling System (I-STEMS) from the programmer's standpoint gets into complex technical decisions regarding programming languages, interprocess communication, parallel and distributed processing, object development, and object encapsulation of legacy software. Perhaps the greatest challenge is the choice of the software building blocks used to develop a large complex system like I-STEMS. Hardware and software environments are still changing very fast. Although it is imperative that system development environments be chosen, emerging technologies can rapidly age such choices. Hence, choices must be made with respect to the anticipated release schedule for the software under development. Since this schedule has not been established, this book only suggests potential choices and focuses on the requirements of the system from the perspective of a programmer.

17.2 System Design Philosophy

I-STEMS is intended to be a general-purpose dynamic, spatial, ecological modeling system. As such, it must be highly modular, adaptable, and interesting to a broad audience of research institutions and programmers. It will not be financially possible for a single organization to design and develop the entire capability. Hence, modularity is an essential requirement.

I-STEMS must allow existing simulation software to be adopted and adapted. As described above, I-STEMS seeks to address the need for land simulation models to communicate with one another. While it may be seductive to imagine the design and development of all new software that can make use of the latest advances in computer hardware and software products and theory, it is essential that I-STEMS developers focus limited I-STEMS resources on techniques that will use existing software. At the expense of simulation efficiency, using existing software avoids the cost of

reproducing, debugging, and supporting replacement software and allows the experts across the modeling and simulation community the opportunity to participate in I-STEMS with minimal investment.

17.3 System Overview

There must be a heart to I-STEMS that provides the glue or focus for the system. The glue will be the underlying submodel intercommunication standards and language. To some extent, it will also be a core set of system viewers and controllers. An overview of the I-STEMS design is captured in Figure 17.1. The three large boxes represent different computers—potentially heterogeneous. Within each computer are a number of different processes connected by data exchange busses. All of the software components operating together represent a simulation model. The ovals represent submodels that a model developer has assembled from a library of modules to address the modeling needs of a land manager.

The solid ovals in Figure 17.1 represent three different submodels that may be legacy software models that may have originally operated as stand-alone programs. Each of these submodels is encapsulated as an I-STEMS object (represented by open ovals). Each object can run as a separate process on a particular computer (or network of computers or multiple central processing units [CPUs] within a single computer). The encapsulation provides standard communication channels for requesting information

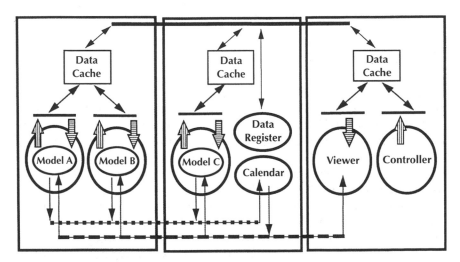

FIGURE 17.1. Overview, ○ I-STEMS object (program); □ Computer; ○ Simulation (sub)-model; ⇑ Data request bject; ⇓ Data request to object

from other submodels (thick arrows with vertical lines) and for responding to such requests initiated from other objects (thick arrows with horizontal lines). Notice the variety of I-STEMS objects (open ovals). Each communicates with other system objects with a standard set of protocols over a limited number of "channels."

Three "channels" are suggested in Figure 17.1. The topmost bar, above the "data cache" objects in the diagram, represents exchange of data among objects as mediated by the data caches. The bottommost bars provide communication between the timekeeper and the systems model objects.

17.4 Subsystem Encapsulation

Each modeling capability added to I-STEMS will conform to strict appearance standards. A design requirement allows direct communication of an encapsulated model component with only the locally running data cache. Hence, any information provided by other model objects running within the same simulation must be provided in standard formats in response to standard requests. This approach makes it possible to add new simulation model components to I-STEMS without a requirement for reprogramming existing components to recognize the new one.

The steps required for subsystem encapsulation of an existing stand-alone simulation model involve:

• Separation of the model from the data and user interface
• Connection of the model to standard I-STEMS encapsulation specifications
• Development of a simulation to test the new model

Before developing these steps, it must be reaffirmed that the best group or person to perform the above steps is the original author of the stand-alone simulation model. Doing so typically minimizes the development costs and helps ensure a link to future developments on the original system. Finally, it minimizes later debugging problems and greatly decreases debugging costs. The core I-STEMS development team will be well advised to contract out most design and development of I-STEMS model components. The core team should focus on the development, enhancement, and maintenance of the core system components described elsewhere in this chapter.

The first step in the list above is to separate the actual modeling code from any data and user interface. An I-STEMS model object interfaces with the rest of the world through communications with its associated data cache. Hence, all communications between any given model and data sources, other models, and humans are accomplished through the data cache. Thus, the actual model must be isolated from all of its communications. Requests for data and the ability to respond to data calls must then be connected to standard I-STEMS model encapsulation routines.

Encapsulation routines will provide a variety of functions that are described below. The actual specification of how these capabilities will be realized is not part of this book. A variety of implementation details are possible and will be the responsibility of an I-STEMS development team. The required functionality is covered here.

- **Register model with local data cache.** Encapsulated I-STEMS submodels will run as separate processes. At start-up time the submodel will be provided with its local cache and with how it communicates with the master clock. At start-up, the model is required to register itself with the local cache.
- **Identify data inputs that will be required at start-up.** Typically, at start-up the submodel identifies to the local cache four types of information. The first type is the information that will be required to initialize the model. The second type is the data that will be required during a simulation. The third type is information that this model can provide at start-up. The fourth type is output that the model can provide during a simulation.
- **Monitor simulation clock.** Another start-up action is to establish communication with the system simulation clock. This is followed by monitoring the clock through the simulation run(s).
- **Register times with clock (announce and wait).** In addition to monitoring the simulation clock, each simulation model will be optionally able to communicate two types of messages to the clock. First, the model can tell the clock to transmit the time at a particular simulation time and then wait for a go-ahead message from model. Second, a go-ahead message can be sent. Events can be scheduled by telling the clock to transmit that time when the scheduled time is met. Those events that must be completed synchronously return a go-ahead upon completion of the event. Asynchronous events will be completed after first sending the go-ahead to the clock. Semisynchronous events can send a second schedule time followed by the go-ahead. This second time represents the time at which the semisynchronous event must be completed. The submodel sends a go-ahead when the event in progress completes.
- **Receive initialization data.** The beginning of a simulation is a unique event. The state of the system being simulated must be loaded into the system. The operation of a simulation involves three basic phases. First, the I-STEMS core simulation software (the main system clock and data register) is started. This software runs on a single machine in a network and is associated with set communication channels. Data caches are also started on each machine participating in the simulation. These caches establish communications with each other and with the data register. Second, the various models, viewers, and controllers are initialized. They communicate their data requirements and offerings to their associated data caches that share this information with the data register. Third, and

finally, after all the data requirements are met, a simulation can be started. Starting involves moving all of the initialization data from data providing objects and then starting the calendar.

- **Request and receive data.** During simulations, data will be moved back and forth among the different models. Each model will request and receive data.
- **Receive and respond to data requests.** Data requests will make their way to the models that supply the data. Each model must accept and respond to these requests.
- **Reset.** Each model must respond appropriately to a reset signal. At this signal, each model will reinitialize itself so that it is in the identical state it was in at first start-up.

As described above, data are being moved among the different models that are running as separate programs. A standard set of data formats will be used to transmit this system state information. These will include:

- A bounded raster of data—integer, floating point, null
- State at a given point—particular piece of information at a particular point
- State within a radius at a given point—returns the average value for a given piece of information
- Others

All data will also be associated with units. Data requests and responses must match units. Unit conversion will be accomplished, when needed, in the process of moving data from the data cache to the requesting submodel.

17.5 Data Cache Objects and the Data Register

I-STEMS submodels do not view the external world as a set of objects, but rather as a set of available information. Each object operates with the "belief" that it is the center of the known universe and is surrounded with information that it can probe. It also responds to information requests, but is unaware of where the requests originate. That external world of information and information requests from the world are mediated by a data cache object. One data cache object is running on each participating computer and is known to each I-STEMS object running on that computer. Because I-STEMS objects do not see other I-STEMS objects directly, the complexity in communicating information is minimized.

The data cache running on any one machine communicates with all other data caches running on all other machines cooperating in a particular I-STEMS simulation. Each cache provides the following functionality:

- Maintain information about data managed by its I-STEMS objects. This information includes the format of the data (raster map, vector map, parameter, value, error measure, and lifetime of the data). This information can be fetched from other data caches when needed and stored locally. It may be provided to its objects when requested. If shared memory is available, it might instead provide the memory location of the data to the objects.

- Map of where each possible data requirement for its I-STEMS objects can be located.

The data cache also communicates with an I-STEMS simulation data register. The data register is associated with a set of "master controls" and manages the location and type of all available data that will be generated by a set of I-STEMS objects. Data caches query this register to find out where each required data type can be found on the network of I-STEMS objects.

17.6 Simulation Timekeeper

I-STEMS submodel objects must run in synchronous simulation time. It is always presumed that the various submodels in an I-STEMS simulation require current information from other submodels. Hence, it is important that each submodel remain synchronous with a central clock keeping simulation time. This is the "calendar" object in Figure 17.1; it accepts requests from the simulation model objects. These objects essentially schedule themselves for updates or actions that they must perform at a later time. For example, a hydrologic simulation object might schedule itself for a full update at a particular time. When the calendar reaches that time it alerts the scheduled object.

17.7 Viewers and Controllers

The right side of Figure 17.1 shows a viewer and a controller. As discussed previously in the section on subsystem encapsulation, the process of converting an existing stand-alone model into an I-STEMS submodel first involved isolating the model from the data and user interface. In I-STEMS, the user interface is replaced with special submodels that, internally, interact with computer peripherals, including monitors, keyboards, and data storage devices. Because these submodels communicate with other components of a complete simulation model via the data caches, the submodels themselves know nothing of the user interface and visualization details.

It is expected that a number of viewers and controllers will be developed to allow appropriate user feedback and input. Some of those expected might be described as follows:

- **Strip chart viewer.** The user will associate one or more streams of data acquired by regularly querying information in submodels. An interactive version will allow a user to peruse the different data available from the submodels and dynamically select the data they wish to track.

- **Error message monitor.** Submodels may generate error messages that can be captured and displayed.
- **Two-dimensional (2D) map.** Mapped information can be dynamically extracted from a submodel and displayed. Various levels of cartographic information, such as grids, overlays, labels, and coordinates, may be optionally displayed.
- **Movie.** A variant of the 2D map, this viewer will allow the history of the simulation to be viewed up to the current simulation time.
- **State monitor.** Some submodels will simulate the state of some discrete watershed entity. The internal state and external environment of these entities may be accessed and viewed.
- **Output capture.** Each of the above monitors may optionally provide the ability to save the data or the images to files for postprocessing. Additionally, some viewers will need to do little more than allow the user to select available data for dynamic storage without rendering the data during a simulation.
- **Person-in-the-loop controller.** This allows a person to simulate the behavior of some watershed entity with a "flight simulator" interface. For example, the behavior of an animal might be provided dynamically by a scientist familiar with that animal. Some submodels might allow a controller to adjust internal parameters, thereby allowing a population to artificially recover, the weather to change, an infestation to begin, or zoning legislation to change. It is likely that some full-simulation models will be developed to do simulations based only on the dynamic input of a number of users. This type of a gaming environment can become very important in exploring alternative approaches to land management.

There will be other viewers and controllers and a number of competing versions of many. I-STEMS must be an open system that can be adopted by a wide variety of research laboratories.

17.8 Implementation Approaches

Creation of an integrated spatiotemporal ecological modeling system can be accomplished with hardware and software technologies currently available. The biggest challenge is amassing sufficient interest in one (or a consortium) of organizations to pull together the first prototype. This is an organizational and leadership challenge. We will first look at alternative technical approaches and then explore potential management approaches.

Before developing any software, it is critical that management target its intended audience. Two critical questions must be answered. First, "Who will be the intended, or target, user community?" Planners? Scientists?

Regional offices with large staffs? Local offices staffed with one or two people? City planning offices? Agricultural planning offices? Second, and easily overlooked, during what years is the system expected to be viable? A system created to be useful for a single user to complete a study next year is much different from a system designed to be viable over a decade or more.

Management of an I-STEMS development project will face several critical challenges. Funding and collaboration are the most important and are inseparable. Collaboration is important partly because it provides a broader funding base. More importantly, however, the goal of I-STEMS is to integrate the best models of scientists into excellent models for making management decisions. Because the models on which I-STEMS integrated models will be based are the end result of significant funding and intellectual effort, it is important to have the original development teams involved in the creation of I-STEMS simulation modules. I-STEMS management will be advised to hold workshops and clinics early and often to maximize the buy-in from the broadest audience possible.

Although there should be significant participation from a large community, I-STEMS must be associated with a small, talented, and dedicated programming staff. This staff must remain consistently funded and remain relatively unchanged throughout the critical first years of development. It will be the responsibility of this team to integrate the good ideas of the community and then return to the community a system with which it will be happy to work.

Conclusions

Chapter 10 discussed three general approaches for applying simulation modeling technologies to watershed management. The first was to use the scientific community as needed to report on the results of their complex models. This approach can be readily used today, although the cost can be significant. The second approach was to use complex scientific models to parameterize much simpler management-oriented decision support systems. Again, this approach has been used successfully for generations, beginning with books and journals authored by scientists for natural resource managers. Scientists collect and analyze information about the world around us and publish conclusions about that world in a condensed form so managers of our world resources might make more informed decisions. More recently, scientists who generate information and conclusions have begun to use the Internet as not only an alternative medium for communicating such knowledge, but also as a medium that allows managers direct and easy access to software that captures the cause–effect relationships in our natural, human, and economic environment. The third approach explored in Chapter 10 was the integration of scientific models into multidisciplinary, management-oriented, spatially explicit simulation models. This approach shows promise for the future.

In Part III of this book we explored ideas that generically define a next-generation management-oriented watershed simulation modeling environment. In summary, the key ideas are:

- A common simulation modeling framework is defined—providing communication links among simulation modules.
- Existing simulation models become modules within the more comprehensive system.
- User interfaces are developed as simulation modules—using the same timing and information exchange channels.
- Libraries of modules are developed, giving managers increased opportunity to accurately map the land/water system into a simulation model.

A number of integrating environments similar to the Integrated Spatio-Temporal Ecological Modeling System (I-STEMS) model already exist, including the Spatial Modeling Environment (SME), the Dynamic Interactive Architecture System (DIAS), the High Level Architecture (HLA), and the Modular Modeling System (MMS). As these systems mature and develop, growing libraries of modules associated with each will emerge. In the meantime, new systems based on new technologies like the extensible markup language (XML), new distributed processing software, and new Internet interfaces and graphics will emerge.

There is no need to wait for the future; where software technology is involved, the future will always hold many more promises than the present holds. Current simulation modeling technologies are powerful, applicable, and available to the natural resource manager. Geographic information systems (GISs) provided virtually the same analysis technologies in 1980 as they do today. In the intervening time, costs for GIS data development and access have dramatically decreased, many people have become educated on GIS capabilities, and faster computers have improved analysis times and dramatically improved user interfaces. Natural resource simulation modeling completes the GIS phenomenon in that it matches understandings of the processes governing our world with the GIS-captured current and historical state of our environment. However, like the GIS of the 1980s, simulation modeling must overcome issues of standardization, costs of model development, and needs for manager-oriented user interfaces that allow the manager to focus efforts on addressing the issues at hand.

References

Adler, R. W., et al. (1993). *The Clean Water Act Twenty Years Later*. Washington, DC, Island Press.

Allen, T. F. H. and T. B. Starr (1982). *Hierarchy*. Chicago, University of Chicago Press.

Andrewartha, H. G. and L. C. Birch (1954). *The Distribution and Abundance of Animals*. Chicago, University of Chicago Press.

Beasley, D. B. and L. F. Huggins (1982). *ANSWERS (Areal Nonpoint Source Watershed Environmental Response Simulation) User's Manual*. Chicago, U.S. Environmental Protection Agency.

Belanger, R., et al. (1989). *ModSim: A Language for Object-Oriented Simulation*. La Jolla, CA, CACI Products Company.

Bodelson, K. and E. Butler-Villa (1995). Santa Fe Institute. http://www.santafe.edu, Internet URL.

Botkin, D. B., et al. (1972). "Some ecological consequences of a computer model of forest growth." *Journal of Ecology* **60**: 849–872.

Botkin, D. B. (1977). Life and Death in a Forest: The Computer as an Aid to Understanding. In C. A. S. Hall and J. W. Day, eds. *Ecosystem Modeling in Theory and Practice: An Introduction with Case Studies*. New York, Wiley, 213–233.

Buckley, D. J., et al. (1993). The Ecosystem Management Model Project: Integrating ecosystem simulation modelling and ARC/INFO in the Canadian Parks Service. *Second International Conference/Workshop on Integrating Geographic Information Systems and Environmental Modeling, Breckenridge, CO*. National Center for Geographic Information and Analysis.

Caswell, H. (1976). "Community structure: A neutral model analysis." *Ecological Monographs* **46**: 327–354.

Caswell, H. (1978). "Predator mediated coexistence: A non-equilibrium model." *American Naturalist* **112**: 127–154.

Chesson, P. L. and T. J. Case (1986). Overview: Nonequilibrium community theories: Chance, variability, history, and coexistence. In J. Diamond and T. J. Case, eds. *Community Ecology*. New York, Harper and Row, 229–239.

Clarke, K. C., et al. (1993). Refining a cellular automaton model of wildfire propagation and extinction. *Second International Conference/Workshop on Integrating Geographic Information Systems and Environmental Modeling, Breckenridge, CO*. National Center for Geographic Information and Analysis.

Clements, F. E. (1936). "Nature and structure of the climax." *Journal of Ecology* **24**: 252–284.

Costanza, R., et al. (1986). Modeling spatial and temporal succession in the Atchafalaya/Terrebonne marsh/estuarine complex in South Louisiana. *Estuarine Variability*. (Douglas A. Wolfe, Ed.) New York, Academic Press, 387–404.

Costanza, R., et al. (1990). "Modeling coastal landscape dynamics." *BioScience* **40**: 81–98.

Costanza, R., et al. (1992). The Everglades Landscape Model (ELM): Summary Report of Task 1, Model Feasibility Assessment. (Project report). University of Maryland, Center for Environmental and Estuarine Studies, Chesapeake Biological Laboratory.

Costanza, R., et al. (1993). Development of the Patuxent Landscape Model (PLM). (Project report). University of Maryland, Center for Environmental and Estuarine Studies, Chesapeake Biological Laboratory.

Costanza, R. and T. Maxwell (1991). "Spatial ecosystem modelling using parallel processors." *Ecological Modelling* **58**: 159–183.

Cronshey, R. G., et al. (1993). GIS–water quality computer model interface: A prototype. *Second International Conference/Workshop on Integrating Geographic Information Systems and Environmental Modeling, Breckenridge, CO*. National Center for Geographic Information and Analysis.

Cuddy, S. M., J. R. Davis, and P. A. Whigham (1993). An examination of integrating time and space in an environmental modelling system. *Second International Conference/Workshop on Integrating Geographic Information Systems and Environmental Modeling, Breckenridge, CO*. National Center for Geographic Information and Analysis.

D'Agnese, F. A., A. K. Turner, and C. C. Faunt (1993). Using geoscientific information systems for three-dimensional regional ground-water flow modeling in the Death Valley region, Nevada and California. *Second International Conference/Workshop on Integrating Geographic Information Systems and Environmental Modeling, Breckenridge, CO*. National Center for Geographic Information and Analysis.

DeAngelis, D. L., et al. (1998). "Landscape Modeling for Everglades Ecosystem Restoration." *Ecosystems* **1**: 64–65.

DeAngelis, D. L. and J. C. Waterhouse (1987). "Equilibrium and Non Equilibrium Concepts in Ecological Models." *Ecological Monographs* **57**(1): 1–21.

Delcourt, H. R. and P. A. Delcourt (1991). *Quarternary Ecology: A Paleoecological Perspective*. London, Chapman and Hall.

DePinto, J. V., et al. (1993). Development of GEO-WAMS: A modeling support system for integrating GIS with watershed analysis models. *Second International Conference/Workshop on Integrating Geographic Information Systems and Environmental Modeling, Breckenridge, CO*. National Center for Geographic Information and Analysis.

DIAS (1995). The Dynamic Information Architecture System: A High Level Architecture for Modeling and Simulation. Argonne, IL, Decision and Information Sciences Division, Argonne National Laboratory.

DMSO (1996). DoD High Level Architecture, http://www.dmso.mil/projects/hla.

Ficks, B. (1997). *Top 10 Watershed Lessons Learned*. Washington, DC, U.S. Environmental Protection Agency, Office of Water, Office of Wetlands, Oceans, and Watersheds.

Fleming, D. M., et al. (1994). ATLSS: Across-Trophic-Level System Simulation for the Freshwater Wetlands of the Everglades and Big Cypress Swamp, National

Biological Service. (Project report). University of TN, http://atlss.org/atlss.report. final.homestead694.txt.

Frederickson, K. E., et al. (1994). A Geographic Information System/Hydrologic Modeling Graphical User Interface for Flood Prediction and Assessment. Campaign, IL, U.S. Army Corps of Engineers, Construction Engineering Research Laboratories.

Frysinger, S. P., et al. (1993). Environmental decision support systems: An open architecture integrating modeling and GIS. *Second International Conference/ Workshop on Integrating Geographic Information Systems and Environmental Modeling, Breckenridge, CO*. National Center for Geographic Information and Analysis.

Funtowicz, S. O. and J. R. Ravetz (1991). A new scientific methodology for global environmental issues. In R. Costanza, ed. *Ecological Economics: The Science and Management of Sustainability*. New York, Columbia University Press, 137–152.

Gardner, R. H., et al. (1991). Simulation of the scale-dependent effects of landscape boundaries on species persistence and dispersal. In M. M. Holland, P. G. Risser, and R. J. Naiman, eds. *The Role of Landscape Boundaries in the Management and Restoration of Changing Environments*. New York, Chapman and Hall, 76–89.

Gardner, R. H., et al. (1993). Ecological implications of landscape fragmentation. In S. T. A. Pickett and M. J. McDonnell, eds. *Humans as Components of Ecosystems: The Ecology of Subtle Human Effects and Populated Areas*. New York, Springer-Verlag, 208–226.

Gaur, N. and B. Vieux (1992). r.fea. Oklahoma City. Campaign, IL, U.S. Army Corps of Engineers, Construction Engineering Research Laboratory.

Ghosh, A. and G. Rushton (1987). *Spatial Analysis and Location–Allocation Models*. New York, Van Nostrand Reinhold.

Gilpin, M. E. (1990). Extinction of finite metapopulations in correlated environments. In B. Shorrocks and I. R. Swingland, eds. *Living in a Patchy Environment*. Oxford, Oxford University Press, 177–186.

Goncalves, P. P. and P. M. Diogo (1994). Geographic information systems and cellular automata: A new approach to forest fire simulation. *European Geographic Information Systems (EGIS) Conference, Paris, France*. EGIS Foundation, 702–711.

Hannon, B. M. and M. Ruth (1994). *Dynamic Modeling*. New York, Springer-Verlag.

Hannon, B. M. and M. Ruth (1997). *Modeling Dynamic Biological Systems*. New York, Springer-Verlag.

Hanski, I. (1985). "Single-species spatial dynamics may contribute to long-term rarity and commonness." *Ecology* 66: 335–343.

Hanski, I. and M. Gilpin (1991). "Metapopulation dynamics: Brief history and conceptual domain." *Biological Journal of the Linnean Society* 42: 3–16.

Hansson, L. (1991). "Dispersal and connectivity in metapopulations." *Biological Journal of the Linnean Society* 42: 89–103.

Hastings, A. (1980). "Disturbance, coexistence, history and competition for space." *Theoretical Population Biology* 18: 363–373.

Hay, L., L. Knapp, and J. Bromberg (1993). Integrating geographic information systems, scientific visualization systems, statistics, and an orographic precipitation model for a hydro-climatic study of the Gunnison River basin, southwestern Colorado. *Second International Conference/Workshop on Integrating Geographic*

Information Systems and Environmental Modeling, Breckenridge, CO. National Center for Geographic Information and Analysis.

Heathcote, I. W. (1998). *Integrated Watershed Management: Principles and Practice*. New York, Wiley.

Hiebler, D. (1994). The SWARM simulation system and individual-based modeling. *Decision Support 2001: 17th Annual Geographic Information Seminar and the Resource Technology '94 Symposium, Toronto, Ontario, Canada*. (J. M. Power, M. Strom, T. C. Daniel, Eds.). Bethesda, MD, American Society for Photogrammetry and Remote Sensing, 474–494.

Horn, H. S. and R. H. MacArthur (1972). "Competition among fugitive species in a harlequin environment." *Ecology* **53**: 749–752.

Johnson, A. R. (1993). Spatiotemporal Hierarchies in Ecological Theory and Modeling. *Second International Conference/Workshop on Integrating Geographic Information Systems and Environmental Modeling, Breckenridge, CO*. National Center for Geographic Information and Analysis.

Kessell, S. R. (1993). The integration of empirical modeling, dynamic process modeling, visualization and GIS for bushfire decision support in Australia. *Second International Conference/Workshop on Integrating Geographic Information Systems and Environmental Modeling, Breckenridge, CO*. National Center for Geographic Information and Analysis.

Kingsland, S. E. (1985). *Modeling Nature*. Chicago, University of Chicago Press.

Kinsel, W. G. (1980). CREAMS: A Field Scale Model for Chemicals, Runoff, and Erosion from Agricultural Management Systems. Washington, DC, U.S. Department of Agriculture. Conservation Research Report #26.

Kirtland, D., et al. (1994). "An Analysis of Human-Induced Land Transformations in the San Francisco Bay/Sacramento Area." *World Resource Review* **6**: 206–217.

Krummel, J. R., et al. (1993). A technology to analyze spatiotemporal landscape dynamics: Application to Cadiz Township (Wisconsin). *Second International Conference/Workshop on Integrating Geographic Information Systems and Environmental Modeling, Breckenridge, CO*. National Center for Geographic Information and Analysis.

Leavesley, G. (1996). The Modular Modeling System. Denver, CO, U.S. Geological Survey.

Leavesley, G. H., et al. (1995). *The Modular Modeling System—MMS: User's Manual*. Denver, CO, U.S. Geological Survey.

Levin, S. A. (1989). Ecology in theory and application. In S. A. Levin, R. G. Hallam, and L. J. Gross, eds. *Applied Mathematical Ecology*. Berlin, Springer-Verlag, 3–8.

Levins, R. (1969). "Some demographic and genetic consequences of environmental heterogeneity for biological control." *Bulletin of the Entomological Society of America* **15**: 237–240.

Levins, R. and D. Culver (1971). "Regional coexistence of species and competition between rare species." *Proceedings of the National Academy of Sciences of the United States of America* **68**: 1246–1248.

Loucks, O. L. (1970). "Evolution of diversity, efficiency, and community stability." *American Zoologist* **10**: 17–25.

Loucks, O. L., et al. (1985). Gap processes and large-scale disturbances in sand prairies. In S. T. A. Pickett et al., eds. *The Ecology of Natural Disturbance and Patch Dynamics*. New York, Academic, 71–83.

MacArthur, R. H. and E. O. Wilson (1967). *The Theory of Island Biogeography*. Princeton, NJ, Princeton University Press.

MacLennan, B. J. (1990). "Continuous spatial automata." University of TN, Knoxville, Department of Computer Science Technical Report CS-90-121, November 1990.

Maxwell, T. (1995). Distributed Modular Spatial Ecosystem Modeling. http://kabir.umd.edu/SMP/MVD/CO.html, University of Maryland.

May, R. M. (1986). "The search for patterns in the balance of nature: Advances and retreats." *Ecology* **67**: 1115–1126.

McLendon, T., et al. (1998). A Successional Dynamics Simulation Model as a Factor for Determination of Training Carrying Capacity of Military Lands. Champaign, IL. U.S. Army Corps of Engineers, Construction Engineering Research Laboratory.

Minar, N. (1995). SWARM Web Pages. http://www.santafe.edu/projects/swarm/, Santa Fe Institute.

Mitasova, H., et al. (1998). Multidimensional Soil Erosion/Deposition Modeling and Visualization Using GIS. Urbana-Champaign, IL, University of Illinois.

Mladenhoff, D. J., et al. (1993). LANDIS: A spatial model of forest landscape disturbance, succession, and management. *Second International Conference/ Workshop on Integrating Geographic Information Systems and Environmental Modeling, Breckenridge, CO*. National Center for Geographic Information and Analysis.

Nee, S. and R. M. May (1992). "Dynamics of metapopulations: Habitat destruction and competitive coexistence." *Journal of Animal Ecology* **61**: 37–40.

O'Neill, R. V., et al. (1986). *A Hierarchical Concept of Ecosystems*. Princeton, NJ, Princeton University Press.

O'Neill, R. V., et al. (1988). "Indices of landscape pattern." *Landscape Ecology* **1**: 153–162.

O'Neill, R. V., et al. (1989). "A hierarchical framework for the analysis of scale." *Landscape Ecology* **3**: 193–206.

O'Neill, R. V., et al. (1992). "A hierarchical neutral model for landscape analysis." *Landscape Ecology* **7**: 55–61.

Oppenheim, N. and R. Oppenheim (1995). *Urban Travel Demand Modeling: From Individual Choices to General Equilibrium*. New York, Wiley.

Overton, W. S. (1977). A Strategy for Model Construction. In C. A. S. Hall and J. W. Day, eds. *Ecosystem Modeling in Theory and Practice: An Introduction with Case Studies*. New York, Wiley, 49–73.

Owen, S. J., et al. (1996). "A Comprehensive Modeling Environment for the Simulation of Groundwater Flow and Transport." *Engineering with Computers* **12**: 235–242.

Pickett, S. T. A., et al. (1989). "The ecological concept of disturbance and its expression at various hierarchical levels." *OIKOS* **54**: 129–136.

Posavac, E. J. and R. G. Carey (1989). *Program Evaluation: Methods and Case Studies*. Englewood Cliffs, NJ, Prentice-Hall.

Price, D. L., et al. (1997). The Army's Land Based Carrying Capacity. Champaign, IL, U.S. Army Corps of Engineers, Construction Engineering Research Laboratory.

Ran, B. and D. E. Boyce (1996). *Modeling Dynamic Transportation Networks: An Intelligent Transportation System Oriented Approach*. New York, Springer-Verlag.

Ray, T. (1994a). Tierra Simulator V4.1. http://alife.santafe.edu/pub/SOFTWARE/ Tierra, Sana Fe Institute.

Ray, T. S. (1994b). "An evolutionary approach to synthetic biology: Zen and the art of creating life." *Artificial Life* **1**: 195–226.

Reice, S. R. (1994). "Nonequilibrium determinants of biological community structure." *American Scientist* **82**: 424–435.

Rewerts, C. C. and B. A. Engel (1991). *ANSWERS on GRASS: Integrating a watershed simulation with a GIS*. St. Joseph, MI, American Society of Agricultural Engineers.

Risenhoover, K. L. (1997). Deer Management Simulator. College Station, TX, Texas A&M University.

Risenhoover, K. L., et al. (1997). A spatially-explicit modelling environment for evaluating deer management strategies. In W. J. McShea, H. B. Underwood, and J. Rappole, eds. *The Science of Overabundance in Deer Ecology and Population Management*. Washington, DC, Smithsonian Institution Press, 366–379.

Rittel, H. W. J. and M. M. Webber (1973). "Dilemmas in a General Theory of Planning." *Policy Sciences* **4**: 155–169.

Robinson, G. R., et al. (1992). "Diverse and contrasting effects of habitat fragmentation." *Science* **257**: 524–525.

Rothermel, R. C. (1972). *A Mathematical Model for Predicting Fire Spread Rate and Intensity in Wildland Fuels*. USDA Forest Service Research Paper. Washington, DC, U.S. Department of Agriculture, Forest Service.

Saaty, R. W. (1987). "The Analytic Hierarchy Process—What it is and how it is used." *Mathematical Modeling* **9**: 161–176.

Saghafian, B. (1993). Implementation of a distributed hydrological model within Geographical Resources Analysis Support System (GRASS). *Second International Conference/Workshop on Integrating Geographic Information Systems and Environmental Modeling, Breckenridge, CO*. National Center for Geographic Information and Analysis.

Singh, V. P., ed. (1995). *Computer Models of Watershed Hydrology*. Highlands Ranch, CO, Water Resources Publications.

Slatkin, M. (1974). "Competition and regional coexistence." *Ecology* **55**: 128–134.

Srinivasan, R. (1992). Spatial Decision Support System for Assessing Agricultural Non-Point Source Pollution Using GIS. (Ph.D. dissertation) Purdue Univ, West Lafayette, IN.

Stommel, H. (1963). "Varieties of oceanographic experience." *Science* **139**: 572–576.

Thau, R. S. (1995). MIT Artificial Intelligence Laboratory. http://www.ai.mit.edu, Massachusetts Institute of Technology.

Tilman, D. and J. A. Downing (1994). "Biodiversity and stability in grasslands." *Nature* **367**: 363–365.

Tomlin, C. D. (1991). Cartographic modelling. In D. J. Maguire, Michael F. Goodchild, and David W. Rhind, eds. *Geographical Information Systems: Principles and Applications*. London, Longman Scientific, 341–374.

Trame, A.-M., et al. (1997). The Fort Hood Avian Simulation Model: A Dynamic Model of Ecological Influences on Two Endangered Species. Champaign, IL, U.S. Army Corps of Engineers, Construction Engineering Research Laboratories.

Turner, M. G. (1989). "Landscape ecology: The effect of pattern on process." *Annual Review of Ecology and Systematics* **20**: 171–197.

Turner, M. G., et al. (1989). "Predicting across scales: Theory development and testing." *Landscape Ecology*. 3(314): p245.

Urban, D. L., et al. (1987). "Landscape ecology, a hierarchical perspective." *Bio-Science* **37**: 119–127.

Urban, D. L., et al. (1991). "Spatial applications of gap models." *Forest Ecology and Management* **42**: 95–110.

Vieux, B. E., et al. (1993). Integrated GIS and distributed stormwater modeling. *Second International Conference on Geographic Information Systems and Environmental Modeling, Breckenridge, CO*. National Center for Geographic Information Analysis.

Vieux, B. E. and J. Westervelt (1992). Finite element modeling of storm water runoff using GRASS GIS. *Computing in Civil Engineering and Geographic Information Systems Symposium, Dallas, TX*. (Barry J. Goodno & Jeff R. Wright, Eds.) Reston, VA, American Society of Civil Engineers, 712–719.

Walde, S. J. (1991). "Patch dynamics of a phytophagous mite population: Effect of number of subpopulations." *Ecology* **72**: 1591–1598.

Whigham, P. A. and J. R. Davis (1989). Modelling with an integrated GIS/expert system. *Procedings of the ESRI User Conference, Palm Springs, CA*. Redlands, CA, Environmental Systems Research Institute.

Wiens, J. A., et al. (1986). Overview: The importance of spatial and temporal scale in ecological investigations. In J. Diamond and T. J. Case, eds. *Community Ecology*. New York, Harper and Row, 145–153.

Wischmeier, W. H. and D. D. Smith (1978). *Predicting Rainfall Erosion Losses— A Guide to Conservation Planning*. Washington, DC, U.S. Department of Agriculture.

Wu, J. (1994). "A spatial patch dynamic modeling approach to pattern and process in an annual grassland." *Ecological Monographs* **64**: 447–464.

Wu, J., et al. (1993). "Effects of patch connectivity and arrangement on animal metapopulation dynamics: A simulation study." *Ecological Modelling* **65**: 221–254.

Wu, J. and O. L. Loucks (1995). "From balance of nature to hierarchical patch dynamics: a paradigm shift in ecology. The Quarterly Review of Biology, **72**: 439–466.

Yaeger, L. (1993). Computational Genetics, Physiology, Metabolism, Neural Systems, Learning, Vision, and Behavior or PolyWorld: Life in a New Context. Cupertino, CA, Apple Computer, Inc.

Young, R. A., et al. (1989). "AGNPS: A nonpoint-source pollution model for evaluating agricultural watersheds." *Journal of Soil and Water Conservation* **44**: 168–173.

Appendix

Indexes of Watershed and Agriculture Land Models

http://www.agralin.nl/camase/
   CAMASE, a Concerted Action for the development and testing of quan
   titative Methods for research on Agricultural Systems and the Environment
ftp://ftp.epa.gov/epa_ceam/wwwhtml/products.htm
   USEPA's Center for Exposure Assessment Modeling
http://www.cee.odu.edu/cee/model/
   Old Dominion University's Civil/Environmental Model Library
http://dino.wiz.uni-kassel.de/ecobas.html
   Register of Ecological Models—WWW Server for Ecological Modelling
   at the University of Kassel
http://hydromodel.com/duan/hydrology/
   Pointers to lots of hydrological modeling resources on the WWW
http://owww.cecer.army.mil/ll/landsimsurvey/homepage.html
   U.S. Army Corps of Engineers database of landscape/watershed models
and modeling environments
http://www.wcc.nrcs.usda.gov/water/quality/common/h2oqual.html
   NRCS reported "Water, Field Scale and Watershed Scale Computer Models,
   Field and/or Point Assessment Tools, and Tools Under Development"
http://www.scisoftware.com/products/prod_alpha/prod_alpha.html
   Alphabetical listing of many models offered commercially by the Scien
   tific Software Group
http://water.usgs.gov/software
   USGS database of public domain models—many compiled
http://www.waterengr.com/
   Watershed Resources Consulting Services

Literature Databases

http://www.nal.usda.gov/ttic/tektran/
   The Agriculture Research Service's Technology Transfer Automated
   Retrieval System

Model Comparision

http://web.aces.uiuc.edu/sriit/watershed/
  Illinois Watershed Management Clearinghouse
http://www.wcc.nrcs.usda.gov/water/quality/common/swf.html
  Field-scale water quality models
http://tsc.wes.army.mil/downloadtracking/DownloadData.asp?PID=75
  Review of soil erosion models

Index